Environmental Studies

THE AUTHOR

Dr. (Mrs.) Rajesh Dhankhar is M.Sc., Ph.D., now working as reader in the Department of Biosciences, M.D. University, Rohtak. She has obtained her degree from M.D. University. She has been teaching about last 14 years in biosciences department, ecology and environmental sciences. She has been actively engaged in researches from last 16 years on ecotoxicology. She has attended various national and international seminars. She has a large number of publications in national and international journals. She has guided several Ph.D. students. Recently she has been involved additionally in teaching IGNOU, engineering and B. Pharmacy. She is also member of various environmental bodies in India.

Environmental Studies

As Per UG, PG and NET Syllabus

Dr. (Mrs.) Rajesh Dhankhar
Reader
Department of Bioscience, M.D. University,
Rohtak – 124 001 (Haryana)

2025
Daya Publishing House®
A Division of
Astral International Pvt. Ltd.
New Delhi - 110 002

First Published, 2006
Reprinted 2020, 2025

ISBN 978-81-7035-433-8 (PB)

Publisher's note:

Every possible effort has been made to ensure that the information contained in this book is accurate at the time of going to press, and the publisher and author cannot accept responsibility for any errors or omissions, however caused. No responsibility for loss or damage occasioned to any person acting, or refraining from action, as a result of the material in this publication can be accepted by the editor, the publisher or the author. The Publisher is not associated with any product or vendor mentioned in the book. The contents of this work are intended to further general scientific research, understanding and discussion only. Readers should consult with a specialist where appropriate.

Every effort has been made to trace the owners of copyright material used in this book, if any. The author and the publisher will be grateful for any omission brought to their notice for acknowledgement in the future editions of the book.

Published by : **Daya Publishing House®**
A Division of
Astral International Pvt. Ltd.
– ISO 9001:2015 Certified Company –
4736/23, Ansari Road, Darya Ganj,
New Delhi - 110 002
Phone: 011-4354 9197, 2327 8134
Fax: +91-11-2324 3060
E-mail: info@astralint.com
Website: www.astralint.com

Digitally Printed at : **Replika Press Pvt. Ltd.**

Preface

Rapid Industrialisation and exponential growth of human population has resulted into a unsustainable development of environment that threatens the life support system, comprising the basic resources like water, soil, air, fauna and flora. Conservation of these natural resources like forest, water and mineral etc. is a primary concern of human race. Developmental policies should be framed that are economically viable and leads to no environmental destruction. To accomplish this, it is necessary to provide a scientific and sustainable development in the area of environment, which is inherently multidisciplinary in nature, involving a number of areas of natural resources, agriculture, medicines, pollution, engineering, law, educational and socio-economic.

The Supreme Court of India has given orders to make all curriculum with a compulsory subject of environment in undergraduate classes.

The objective of this book is to provide multidisciplinary study of a wide range of environmental related issues. It consists of seven units of theory and one unit is of field experiments. The first seven units emphasized on multidisciplinary nature of environment and about public awareness. The other six units consists of natural resources namely Forest resources, Water resources, Mineral resources, Food resources, Energy resources, Land resources, Ecosystems, Biodiversity, Pollution and Social issues in relation to environment.

This book is not only for undergraduate students but also for postgraduate students and those appearing in competitive examination like National Entrance Test (NET).

I express my sincere thanks to Dr. Naseeb Singh Gill, Computer Science Department, M.D. University, my students Ms. Sushila Sangwar, Ms. Nitant Gaur and Mr. Sunil Chikara for providing latest information from various sources. I also thanks to Daya Publishing House, Delhi for the edition of this book.

Rajesh Dhankhar

Contents

Core Module Syllabus for Environmental Studies for Undergraduate Courses of All Branches of Higher Education

Unit 1: The Multidisciplinary Nature of Environmental Studies

Definition, scope and importance (2 Lectures)

Need for public awareness

Unit 2: Natural Resources

Renewable and non-renewable resources:

Natural resources and associated problems

(*a*) **Forest resources**: Use and over-exploitation, deforestation, case studies. Timber extraction, mining, dams and their effects on forests and tribal people.

(*b*) **Water resources**: Use and over-utilization of surface and ground water, floods, drought, conflicts over water, dams benefits and problems.

(*c*) **Mineral resources**: Use and exploitation, environmental effects of extracting and using mineral resources, case studies.

(*d*) **Food resources**: World food problems, changes caused by agriculture and over-grazing, effects of modern agriculture, fertilizer-pesticide problems, water logging, salinity, case studies.

(*e*) **Energy resources**: Growing energy needs, renewable and non-renewable energy sources, use of alternate energy sources, case studies.

(*f*) **Land resources**: Land as a resource, land degradation, man induced landslides, soil erosion and desertification.

- ☆ Role of an individual in conservation of natural resources.
- ☆ Equitable use of resources for sustainable life styles.

(8 Lectures)

Unit 3: Ecosystems

- ☆ Concept of an ecosystem.
- ☆ Structure and function of an ecosystem.
- ☆ Producers, consumers and decomposers.
- ☆ Energy flow in the ecosystem.
- ☆ Ecological succession.
- ☆ Food chains, food webs and ecological pyramids.

- Introduction, types, characteristic features, structure and function of the following ecosystem:

  (*a*) Forest ecosystem

  (*b*) Grassland ecosystem

  (*c*) Desert ecosystem

  (*d*) Aquatic ecosystem (ponds, streams, lakes, rivers, oceans, estuaries).

(6 Lectures)

Unit 4: Biodiversity and its Conservation

- Introduction–Definition: genetic, species and ecosystem diversity.
- Biogeographical classification of India.
- Value of biodiversity: consumptive use, productive use, social, ethical, aesthetic and option values.
- Biodiversity at global, national and local levels.
- India as a mega-diversity nation.
- Hot-spots of biodiversity.
- Threats to biodiversity: habitat loss, poaching of wildlife, man-wildlife conflicts.
- Endangered and endemic species of India.
- Conservation of biodiversity: *In-situ* and *Ex-situ* conservation of biodiversity.

(8 Lectures)

Unit 5: Environmental Pollution

Definition

- Causes, effects and control measures of

  (*a*) Air pollution

  (*b*) Water pollution

  (*c*) Soil pollution

  (*d*) Marine pollution

  (*e*) Noise pollution

  (*f*) Thermal pollution

  (*g*) Nuclear hazards

- Solid waste management: Causes, effects and control measures of urban and industrial wastes.
- Role of an individual in prevention of pollution.
- Pollution case studies.
- Disaster management: floods, earthquake, cyclone and landslides

(8 Lectures)

Unit 6: Social Issues and the Environment

- ☆ From Unsustainable to Sustainable development.
- ☆ Urban problems related to energy.
- ☆ Water conservation, rain water harvesting, watershed management.
- ☆ Resettlement and rehabilitation of people: its problems and concerns. Case studies.
- ☆ Environmental ethics: Issues and possible solutions.
- ☆ Climate change, global warming, acid rain, ozone layer depletion, nuclear accidents and holocaust. Case studies.
- ☆ Wasteland reclamation.
- ☆ Consumerism and waste products.
- ☆ Environment Protection Act.
- ☆ Air (Prevention and Control of Pollution) Act.
- ☆ Water (Prevention and Control of Pollution) Act.
- ☆ Wildlife Protection Act.
- ☆ Forest Conservation Act.
- ☆ Issues involved in enforcement of environmental legislation.
- ☆ Public awareness

(7 Lectures)

Unit 7: Human Population and Environment

- ☆ Population growth, variation among nations.
- ☆ Population explosion–Family Welfare Programme.
- ☆ Environment and human health.
- ☆ Human Rights.
- ☆ Value Education.
- ☆ HIV/AIDS.
- ☆ Women and Child Welfare.
- ☆ Rote of Information Technology in Environment and human health.
- ☆ Case Studies.

(6 Lectures)

Unit 8: Field Work

- ☆ Visit to a local area to document environmental assets–river, forest grassland/hill/mountain.
- ☆ Visit to a local polluted site–Urban/Rural/Industrial/Agricultural.
- ☆ Study of common plants, insects and birds.
- ☆ Study of simple ecosystems–pond, river, hill slopes etc. (Field work Equal to 5 lecture hours)

(5 Lectures)

1

The Multidisciplinary Nature of Environmental Studies

CHAPTER OBJECTIVES

- Definition, Scope and Importance
- Need for Public Awareness

DEFINITION, SCOPE AND IMPORTANCE

The relationship between the organisms and environment is called ecology. The term ecology can be split into — *Eco*-house, *logos* – study. This term was first coined by a German biologist Ernest Haeckel in 1869.

The term environment etymologically means surroundings. So environment is a complex blend of things which surround an organism. It can be defined as the sum total of social, economical, biological, physical or chemical factors which surround the human beings. The definition clearly indicates the sum total of air, water, land and their interactions with the human beings and other living organisms.

Environmental studies is a multidisciplinary subject because it involves various disciplines with a holistic approach. To study environment one should have knowledge inputs from different aspects. Physical and chemical nature of the abiotic components can be understood with the help of geology, chemistry, physics, geography, statistics, mathematics and computer science. All the above subjects act as tools for environmental modelling and management. The biotic components and their interactions with the environment can be understood with the help of biosciences like botany, genetics, microbiology, biochemistry, zoology and biotechnology.

Besides this, various environmental development activities require the knowledge of subjects like education, economics, mass communication, sociology, psychology, and so on. All these subjects act as inputs for dealing with the socio-economic aspects of environment.

For effective management and protection of the environment, one should have the knowledge of environmental law. Professionals like civil engineers, hydraulic engineers and chemical engineers play an important role in protecting the environment by controlling pollution and developing waste treatment technology.

It covers large areas and prospects. Environmental studies mainly emphasizes on industries, mining, biodiversity, conservation of natural resources and their management, environmental pollution and control, population and social issues in relation to development of environment. Actually, environment is important for all living beings. Every aspect of life has a direct relation to environment. Hence, environmental studies can be studied as a specialised subject with an emphasis on many aspects like environmental sciences, environmental engineering, environmental management and environmental technology.

Besides these, there are many global issues like global warming, loss of forests, misuse of energy resources, depletion of ozone layer, loss of biodiversity, increase in population, to name a few. All these factors indirectly effect the environment. Hence, we have to concentrate on these aspects globally. Also, there are many local problems which include industrial effluent, solid waste management, mining and so on. To resolve all these local as well as global environmental problems we will have to make people educated on environment.

Environmental studies are important not only for students but also for house wives and farmers as well, because it deals with issues like safe and clean drinking water, clean air, fertile land, hygienic conditions, nutrient rich food, and sustainable development.

Environmental studies includes the following areas:

1. Natural resources—conservation and management.
2. Ecology and Biodiversity.
3. Environmental pollution and it's control.
4. Social issues in relation to development and environment.
5. Human population and environment.

These are the basic aspects of environmental studies and have a direct relevance on every section of the society. Environmental studies can be highly specialized which may concentrate on technical aspects like environmental science, environmental engineering or environmental management.

NEED FOR PUBLIC AWARENESS

In 1981, an international conference was held on environment which emphasized the need for environmental education at all levels. Bhopal disaster in 1984 focussed the attention on the need of environmental awareness in India. Bhopal gas tragedy was caused due to lack of environmental awareness at all levels–central government, state government, hospitals, doctors, industry and the public.

Hence, environment education is important for the overall development of a country. People must be made aware of the present state of environment–causes and it's remedies. Basic concept and essence of environment should reach the entire population.

The United Nations Conference on Environment and Development was held in Rio de Janerio in 1992 and was popularly known as "Earth Summit". It was followed by World Summit on Sustainable Development at Johannesberg in 2002.

The goals of sustainable development cannot be achieved by any government at its own level until the public has a participatory role in it. Public participation is possible only when the public is aware about ecological and environmental issues. A drive by the government to ban the littering of polythene has not been successful due to the public unawareness on environment.

There is a Chinese proverb saying "If you plan for one year, plant rice, if you plan for 10 years, plant trees and if you plan for 100 years, educate people."

If we want to manage our Earth we will have to make all people environmentally educated. In 1991, Supreme Court of India issued directives to make all curricula environment oriented. Today, everyone talks of environment but only a few have a clear idea about it. Henry D. Thorean rightly says "What is the use of a beautiful house if you don't have a decent Planet to put it on." If we begin today, the environment restoration is expected in the next 40-50 years.

REVIEW QUESTIONS

Short Type Questions

1. Define the term ecology?
2. Who introduced the term ecology?
3. What are the components of environment?
4. What is environmental studies?
5. What are the biotic and abiotic components in an ecosystem?
6. What are the environmental problems in our everyday life?

Descriptive Questions

7. Explain the term environment.
8. Discuss different aspects of environmental studies.

9. How is environmental study important in our everyday life?
10. Discuss the scope of environmental studies.
11. Discuss environmental problems and their effective solutions.
12. Discuss the need for public awareness about environment in the country.

2
Natural Resources

CHAPTER OBJECTIVES

- Introduction
- Forest Resources
- Water Resources
- Mineral Resources
- Food Resources
- Energy Resources
- Land Resources
- Role of an Individual in Conservation of Natural Resources

INTRODUCTION

In broad terms, a resource or natural resource, is an asset available to be used when needed by an organism or group of organisms. In other words, a resource is something useful for humanity, but the concept of utility can change because of technological, economical, and environmental factors.

The sum total of all physical, chemical and social factors which compose the surrounding of man is referred to as environment and each element of these surroundings constitute resource on which man draws in order to develop a better life. Any material which can be transformed in a way that it becomes more valuable and useful can be termed as resource. It is possible to obtain valuable items from any resource.

Ramade (1984) defined a resource as a form of energy or matter which is essential for the functioning of organisms, population and ecosystem. In the case of human beings, a resource, in his words, "is any form of energy or matter essential for fulfilment of physiological, socio-economic and cultural needs, both at the individual level and that of the community".

The functioning of ecosystem or of technological civilisation involves a continuous flow and consumption of energy. In light of this fact, a resource may be defined as one of the various forms of primary source of energy present in nature. The five basic ecological variables are energy, matter, space, time and diversity, sometimes together called natural resource.

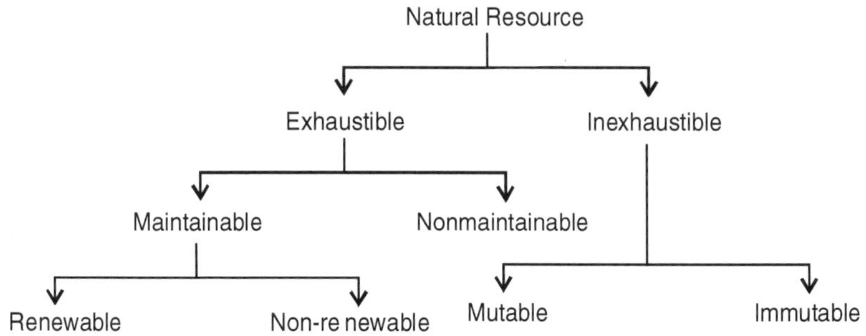

Classification of Natural Resources

Resources may be classified on the basis of the following.

1. **Availability**
   - (i) **Inexhaustible:** These resources are unlimited in quantity and quality. With the passage of time quality may be degraded but not the quantity.
   - (ii) **Exhaustible:** Quantity and quality of these resources may be exhausted soon.
2. **Renewed Capability**
   - (i) **Renewable:** Resources which can be renewed along with their exploitation. For example forest, solar energy.
   - (ii) **Non-renewable:** These cannot be regenerated if exploited on a large scale. For example mineral, water and so on.
3. **Quality**
   - (i) **Immutable:** Quality of these resources are not degraded by human activity. For example tidal power, wind, atomic power.
   - (ii) **Mutable:** Quality may be degraded by man's activity. For example water, hydropower.
4. **Method used**
   - (i) **Maintainable:** The method used for consumption by human being.
   - (ii) **Non-maintainable:** Total quantity static when consumption occurs.
5. **Origin**
   - (i) **Biotic:** (Organic) Biotic resources are obtained from biosphere. For example crops, birds, animals and so on.
   - (ii) **Abiotic:** (Inorganic) Resources composed of non-living matter. For example land, water, minerals.

Anything required by us for living and provided by nature is called a natural resource. These are of two types : Renewable and Non-renewable resources.

Renewable resources–These are inexhaustible and can be regenerated within a given span of time.

Solar energy is a form of renewable resource. It is in the form of heat, wind, water and wood and can be converted into other forms of energy. Examples of renewable resources are grasslands, forests trees, animals, lakes, ground water, fertile soil and fresh air.

Non-renewable resources–These cannot be regenerated. Hence, they need to be used in a judicious manner. These are also known as exhaustible resources because they exist in fixed quantities in the Earth's crust. For example coal, copper, oil and aluminium. Now 80 per cent of the estimated supply has been extracted from earth. Non-renewable resources such as oil, coal and natural gas cannot be recycled and reused. But contrary to this, renewable material resources can be recycled and reused.

FOREST RESOURCES

Forests are one of the most important natural resources on earth. They cover earth like a blanket and provide innumerable material goods and environmental services.

Forests are important for ecological balance of an area. They are essential components of our environment and economy. About 1/3rd of earth's land area is covered with forest. Although natural forest have declined every where in the world, greatest loss has occured in tropical Asia. USSR accounts for fifth and Brazil for seventh of the world's forest.

Uses of Forests Following are the uses of forests:

1. **Check air and noise pollution:** Forests can absorb many toxic gases and thus keep the air pure. Trees can also absorb noise and lower down the noise pollution level.
2. **Reduce wind erosion:** In deserts, forests reduce wind erosion by checking wind velocity.
3. **Reduce soil erosion:** Forests bind the soil particles with root and prevent soil erosion, particularly in hilly areas.
4. **Prevent landslides:** Forests also reduce landslides by binding the soil particles with their roots.
5. **Protection to wildlife:** Forests provide homes to millions of wild animals and plants.
6. **Balancing the gaseous cycles:** Forests balance the carbon dioxide and oxygen ratio in the atmosphere.
7. **Production of oxygen:** The trees produce oxygen by photosynthesis and are also known as earth's lungs.
8. **Reduce global warming:** Trees absorb carbon dioxide and release oxygen during photosynthesis thereby reducing carbon dioxide concentration in the atmosphere.

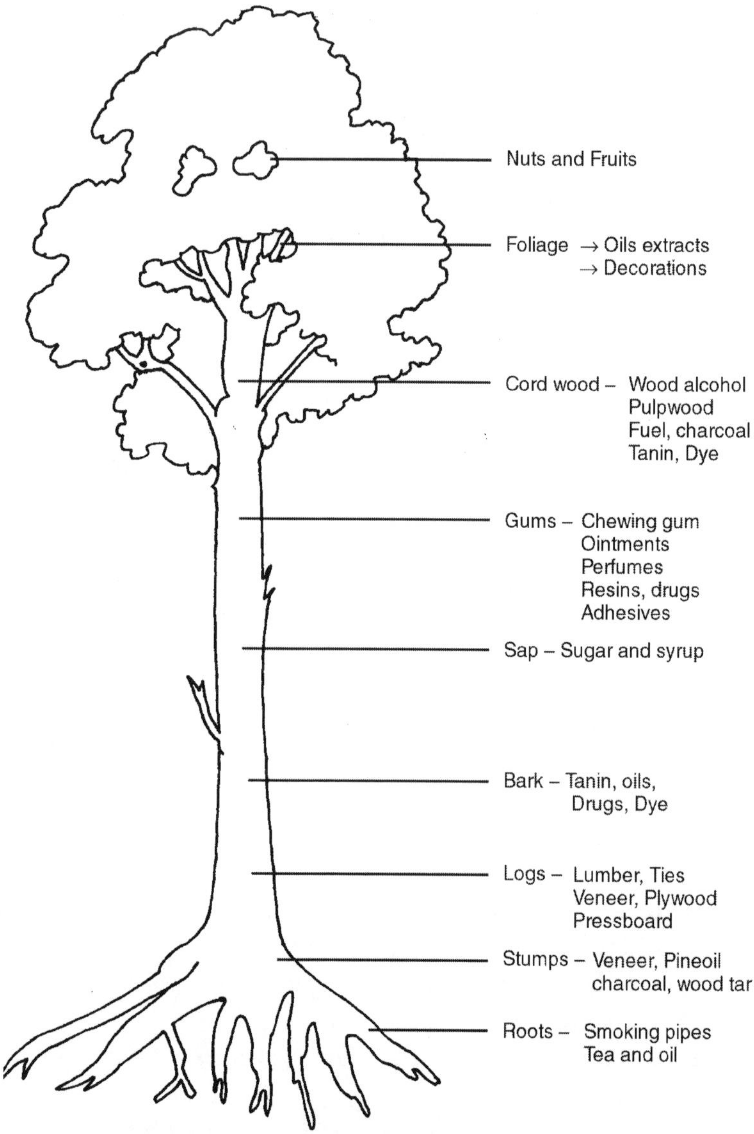

Fig. 2.1 Trees showing useful products from its parts

9. **Regulation of hydrological cycle:** Forested water shed act like giant sponges, absorbing the rainfall, slowing down the runoff and slowly releasing the water for recharge of springs. About 50-80 per cent of the moisture in the air above tropical forests comes from their transpiration and helps in bringing rains.

10. **Reduce flood havoc:** Trees reduce the flood havoc by absorbing the excess water.
11. **Aesthetic uses:** Trees also play a dominant role in the beautification of the environment.
12. **Recreational uses:** With the development of national parks and sanctuaries, trees have a role in recreation too.
13. **Commercial uses:** Trees produce commercial goods. They mainly provide wood which is used as fuel. Wood is also used as a raw material for various industries as pulp, paper, newsprint, board, timber for furniture items. It is used for packing material, matches, sport goods, etc.

Over Exploitation of Forests

Forests have served mankind from the very beginning of its history. Now, with growing population the demand for raw materials like timber, fuel wood has increased and resulted in large-scale wood felling, destruction, road building and clearing of forests. Forests also contribute to the national economy. The international timber trade alone is about US $40 billion per year. But the excessive use of fuel wood and charcoal, expansion of urban, agricultural and industrial areas, overgrazing, deforestation, desertification, have together led to exploitation of forest.

Deforestation

Forest area of the world was about 7,000 million hectares in the year 1900. It reduced to 2890 million hectares in 1975 and further fell down to 2,300 million hectares by 2000. India is poorest in the world as regards to per capita land is concerned. Today, per capita forest land in India is 0.10 hectare as compared to world average of 1 hectare. India is losing about 1.5 million hectares of forest cover each year. India consumes nearly 170 million tons of firewood annually and 10-15 million hectares of forest cover is stripped every year to meet fund requirements. About 1 per cent of the land surface in India turns barren each year due to deforestation. Deforestation is more prominent in tropical than temperate countries.

Causes of deforestation

1. **Explosion of human population:** Overpopulation has increased the requirement of timber, fuel, wood etc. Their consumption in

India has increased up to 300 to 500 million tons in 2001 as compared to just 65 million tons during independence.

2. **Shifting cultivation:** (Jhum) The Jhum or shifting cultivation is a traditional agro-forestry in the north eastern part of India. It involves felling of trees, followed by cultivation of crops for a few years.
3. **Raw materials for industries:** Raw materials like wood, used for industrial processes such as making of crates, boxes, furniture, railway sleepers, pulp industry and so on have put pressure on forests. Plywood requirement is maximum for packing tea and apples.
4. **Dams and hydroelectric projects:** Dams and hydroelectric projects displace large local population and cause great damage to the forests.
5. **Road construction:** Construction of roads along the mountains cover about 30,000 km and cause forest degradation.
6. **Mining operations:** Mining has a severe impact on forest area. Large forest areas are cleared and converted into pastures.
7. **Growing food needs:** Due to growing population, forest covers are converted into permanent agricultural lands for food.
8. **Fires:** Fires destroy full grown forests and damage the wild life.
9. **Pests:** Pests destroy trees by eating up of leaves and boring the shoots.
10. **Overgrazing:** Overgrazing has resulted in large scale degradation of forests.
11. **Weather conditions:** Weather conditions such as storms, frost and heat, destroy the forests.

Consequences of deforestation

The main consequences of deforestation are as follows:

- Deforestation reduces biodiversity and genetic diversity.
- Deforestation reduces rainfall by disturbing hydrological cycle.
- Deforestation increases soil erosion.
- Deforestation disturbs the natural habitat of wild species.
- Deforestation leads to landslides in hilly areas.
- Deforestation induces global climatic change leading to global warming.
- Deforestation threatens indigenous tribal people who depend on forests for their survival.

In newspaper industry, requirement of forest raw material is as following:

Table No. 2.0

Year	Capacity Requirement (Mt)
1981	0.467
1986	0.596
1991	0.770
1996	1.000
2000	1.289

Case Study in Tropical Forests

Tropical forests are considered as non-renewable resources. Once the forest area is cleared, uppermost fertile layer of the soil erodes away. Moisture once captured, is released to the atmosphere and is no longer available, so that the area once covered with forests tends to become dry and desert like. Tropical forests are the sources of rubber, nuts, fruits, wood and fodder. Due to continuous and uninterrupted exploitation, this ecosystem has been damaged badly. A large number of plants and animal species are already extinct and some are in the danger of getting extinct.

Special efforts are being made all over the world not only to preserve flora and fauna but also to regenerate forests and develop wild life and its habitat. Tigers and Indian bustards are examples of endangered species.

Case Study in Deforestation

Deforestation involves clearance of natural forests and plantation of monocultures like Pinus *roxburghie, Eucalyptus Camadulensis* and so on of soil. They distrub the ecosystem by changing biological properties resulting in it's degradation. North East Himalayas of West Khasi hill, some parts of Garhwal, Kumaon and Ladakh are examples of desertification.

Timber extraction Timber extraction involves the cutting of valuable timber like teak and mahogany. It consists of a few large trees per hectare, which are interlocked with each other by means of vines. Road construction near the mountains also causes damage to the forest.

Mining Large scale deforestation due to mining has been reported in Mussorie and Dehradoon Valley. The forest area has declined up to 33 per cent due to mining activities. Mining operations for extracting minerals involve large forest areas. About 480,000 hectare of land is under mining activities in India. Mining involves the removal of vegetation along with underlying soil mantle and overlying rock masses. This leads to destruction in topography and landscapes of that area.

Dams and their effects on forests and tribal people Our late Prime Minister Pt. Jawahar Lal Nehru called dams as "Temples of Modern India" because dams store plenty of rain water for irrigation and generate electricity. But these dams are also responsible for the destruction of large areas of forests. There are about 1550 large dams. Some of them are Bhakra-Nangal, Heerakund, Nagarjuna Sagar, Damodar, Sardar Sarovar, Tehri etc.

Several small dams cause less damage to the environment than a big dam. Big dams have been in sharp focus of various environmental groups all over the world. Construction of dams leads to ecological problems like deforestation and displacement of local people from these habitations.

Effects on forests and tribal people Following are the effects on forests and tribal people:

1. **Submergence of forest land:** A large area of forest land gets submerged during construction of a dam.
2. **Devastation of forests:** This leads to felling and cutting of trees and leads to ecological imbalance in the area.
3. **Prone to floods, droughts and landslides:** The areas where deforestation occurs becomes prone to floods, droughts and landslides.
4. **Loss of soil fertility:** The soil become less fertile due to decrease in decomposition of dead leaves in that particular area.
5. **Loss of biodiversity:** By constructing dams and cutting forest, biodiversity is at great loss as forests are the repositories of invaluable gifts of nature in the form of biodiversity.
6. **Loss of unidentified species:** There are many species which are not known to mankind but have marvelous medicinal and economical value, get lost due to construction of dams.

7. **Effects on tribal people:** A number of villages get submerged under water. These submerged areas were rich in wild life. Tribal people, who use wild flora and fauna as food, hence, are deprived off.
8. **Displacement of tribal People:** Millions of tribal people get displaced from the original area. This displacement affects the tribal culture, their beliefs, myths, rituals, festivals and so on. Tribal people face hardships for the so called development and prosperity of their land.

WATER RESOURCES

Water is a magic resource which sustains life. Water plays a crucial role in biological diversity and in determining the climate of a place. Water is considered a base for economic development of a country.

Water is the main constituent of hydrosphere. It is a renewable resource. About 97 per cent of earth's surface is covered by water. Global resources of water are about 1404 million km^3

Table 2.1: Resources of water in the hydrosphere

Reservoirs	Percentage of total water
Ocean	97.54
Ice	1.81
Ground water	0.63
Lakes and streams	
1. Salt	0.007
2. Fresh	0.009
3. Atmosphere	0.004

Figures account for over 99.9 per cent of the water.

Water Use and Exploitation

About 97 per cent of earth's total supply of water is found in the oceans and remaining 3 per cent is in form of fresh water and distributed unevenly. But over ¾ of water is unavailable for use by plants, human beings and other animals because it is locked up in glaciers, polar ice caps, atmosphere and soil. Only 0.5 per cent of earth's water is available as fresh water in rivers, lakes and as ground water. Human beings depend on water for almost every developmental activity. Water shapes the earth's surface and regulates our environment.

The rapid rise in population and industralization has increased the demand for water. The use of water has increased by 4 to 8 per cent per year. Since 1950, agriculture has been using 70 per cent of the water available to us. Hence, agriculture involves maximum consumption of water in the world. Only 1.1 per cent is used for domestic and municipal supplies and rest of the water is used for various activities like, pharmaceutical, cement, mining, detergent, textile and leather industries. In India we use 93 per cent of the water in agriculture while in Kuwait only 4 per cent is used for watering crops.

100 litres (25 gallons)

100% (Water Resources) → 3.2% (Fresh water) → 0.5% (Available fresh water) → 0.003% (Usable fresh water)

Uses of water are broadly classified into:

- Agricultural uses
- Industrial uses
- Domestic uses
- Aquatic culture e.g. fish culture etc.
- As source of minerals
- As source of energy

There are two common measures of water use

1. **Water withdrawl:** It involves taking of water from ground water or surface water resource.
2. **Water consumption:** It involves water which is taken up but not returned for reuse.

Only 60 per cent of the water withdrawn is consumed because rest is lost through evaporation.

Surface water run off The water which comes through precipitation and does not infiltrate into the ground or does not return to the atmosphere is known as **surface water**. It becomes run-off, when water flows into nearby streams, rivers, lakes, ponds, wetland and reservoirs. The surface water is used for irrigation purpose, industrial use, public use and navigation etc.

Flood Massive run-off of water due to heavy rainfall leads to flooding in the low-lying coastal areas. Human activities like deforestation,

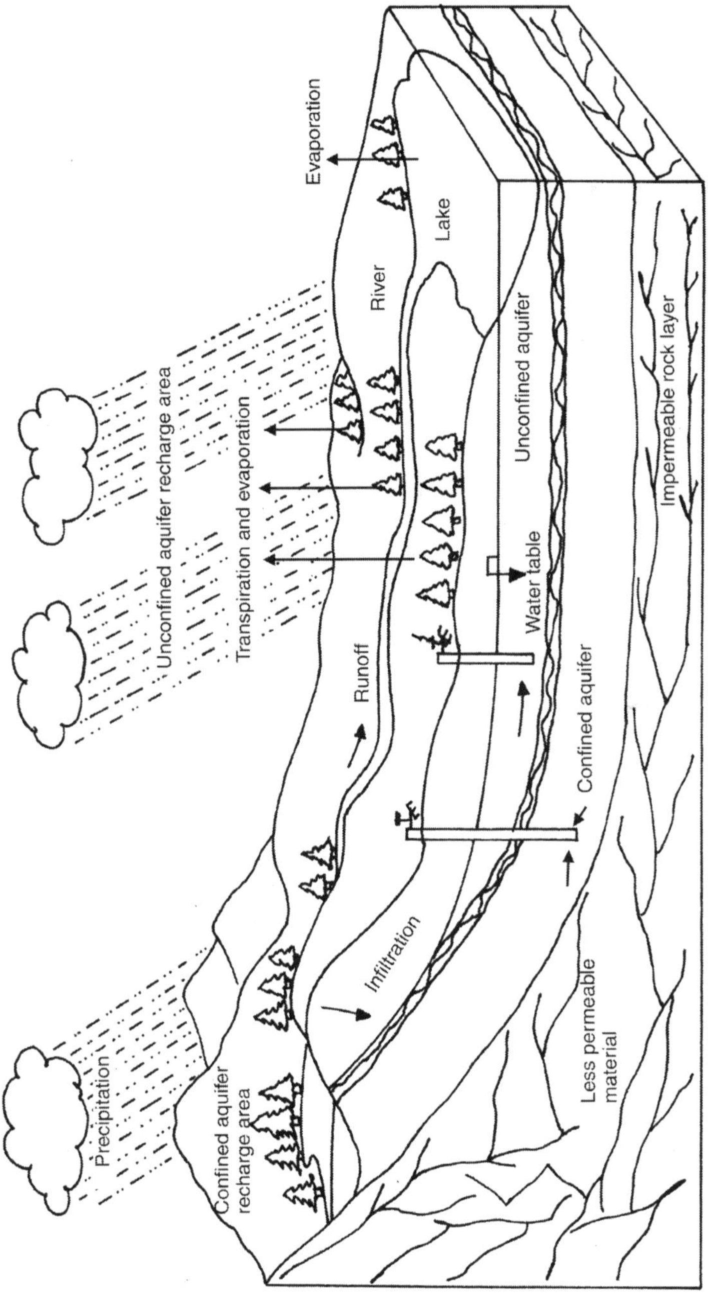

Fig. 2.2 The ground water system

urbanization, cultivation of marginal lands, overgrazing, mining and so on contribute to increase in frequency of floods by removing vegetation which absorbs water. India being a monsoon land, rainfall is confined to 3 to 4 months. So there are number of effective methods for preventing or reducing floods:

- Replanting vegetation in disturbed areas to reduce the run-off.
- To create ponds in urban areas to retain rain water and release it in the rivers.
- Diversion of rainwater towards holding tanks for use by industries.
- Flood prone areas should be purchased by government and converted into parks etc.

Droughts Nearly 80 countries or about 40 per cent of the world population experiences severe droughts. Most of these countries are from Asia and Africa. Reduced precipitation, less annual rainfall and evaporation leads to drought conditions. The drought hit areas have poor land use efficiency and makes life difficult for the population.

Proper understanding of the ecological requirements and natural processes can be useful for controlling the drought conditions. For example, in Karnataka, a project on drought has been undertaken by social forestry with world bank aid. But due to *Eucalyptus* tree plantation, the water table has lowered because euclyptus has high transpiration rate. Factors responsible for drought conditions are overgrazing, deforestation and mining etc. Social forestry and wasteland development are two effective methods to combat the problem of drought. Mixed cropping also helps to optimize the drought conditions.

Ground Water

Some water percolates into the ground through precipitation and it partially fills pores between soil particles and rocks within the upper soil and rock layers of Earth's crust. These layers of Earth's crust are termed as **aquifers** and the water in the layers of Earth crust is called **ground water**.

Out of total fresh water only 9.86 per cent is in the form of ground water. Groundwater is considered to be very pure. But due to contamination by leachates from sanitary landfills, it becomes impure.

Aquifers The layers of Earth crust or rock that is highly permeable and transmits water is called aquifer. Layers of sand and gravel are good aquifers while clay and crystalline rocks have low permeability.

Aquifers can be divided into two types. They are as follows :

1. **Confined aquifers:** It is also known as artesian aquifers. It is formed when groundwater is sandwiched between two impermeable layers of rock or sediments. Recharge of ground water occurs only in those areas where the aquifer intersects the land surface. Sometimes the recharge area is hundreds of kilometers away from the location of the well.
2. **Unconfined aquifer:** It is also known as water table aquifer. It is formed when groundwater collects above the impermeable layers of rock. Recharge occurs by water seeping down from above in the form of rainfall and snow melt. If ground water withdrawal is more than its recharge rate, the sediments in the aquifer get compact and the phenomenon is called **ground subsidence**. Due to this phenomenon large economic losses may occur resulting in sinking of land surface. Structural damage may occur in buildings, fracture in pipes, may reverse the flow of sewers etc.

Conflict Over Water

Unequal distribution of water leads to inter-state and international disputes. These disputes ultimately affect our farmers due to unavailability of water in their fields. Some of the water conflicts are discussed here.

Middle East Jordan, Tigris-Eupharates and Nile are three river basins where water resources are shared by middle east countries. Sudan and Ethopia have more volume of water than Egypt. Similarly there is fight for water among Jordan, Israel and Syria for the Jordan river. Turkey has more water than Syria and Iraq due to development of dam on Tigris–Euphrates.

Indus Basin Indus Water Treaty was established in 1960. In this the Indus, the Chenab and the Jhelaum were allocated to Pakistan while the Ravi, the Beas and the Satluj were allocated to India. The Indus water treaty requires that the three rivers of Pakistan may be used for non-consumptive purposes by India. But due to improving relations between the two countries it is desirable to go for an integrated development of the river basin in a sustainable manner.

Satluj-Yamuna Link and SYL Canal dispute The dispute is between two states, namely, Haryana and Punjab over the sharing of Ravi-Beas waters and Satluj-Yamuna Link. The conflict is that Punjab stakes its claim over Beas, Ravi and Satluj while Haryana stakes claim on Yamuna basin. But now Haryana has faced acute shortage of water, as it has tried to help U.P., Rajasthan and Delhi by signing MOU. The water conflicts can be resolved by greater understanding and objectivity.

DAMS—Advantages and Disadvantages

Advantages

India has many big dams and river valley projects. Big dams are considered to play key role in the development of the country. These are the symbol of national development. The tribal people living in the areas have big hopes for employment after the construction of dams. The big dams have a great potential for economic upliftment and growth. Dams control the flow of rivers and can reduce the danger of flooding, generate electricity, reduce water and power shortage, promote navigation and provide drinking water in remote areas. Rainwater and snow water can be captured by dams and stored in reservoirs associated with dams. Dams provide supply of water for irrigating land. Big dams create reservoirs that can be used for swimming, boating and fishing.

Disadvantages

Big dams have many environment problems and have now become a subject of controversy. The problems can be categorized into two levels, that is, upstream and downstream.

Problems related to upstream are as follows:

- Loss of forests.
- Migration and spawning of fish.
- Displacement of tribal people.
- Loss of non-forest land.
- Water logging near reservoir.
- Siltation of reservoir.
- Seismicity induced by reservoir.
- Growth of weeds.
- Breeding sites for vector borne diseases like malaria.
- Microclimatic changes.

Problems: related to downstream are as follows:

- Micro-climatic changes.
- Reduced flow of nutrients and fresh water into estuaries.
- Loss of soil fertility near the dam.
- Salt water intrusion at the river mouth.
- Flash floods.
- Water logging due to irrigation in excess.
- Salinity due to increase irrigation in excess.

MINERAL RESOURCES

Minerals are non-renewable and exhaustible resources. These are naturally occurring, crystalline solids and inorganic in nature. Minerals have a definite chemical composition and physical properties. Minerals are of two types (1) metallic and (2) non metallic. Mineral is a homogenous, naturally occuring solid inorganic substance with a definable chemical composition and external structure characterised by an orderly arrangement of atoms, ions or molecules in a lattice.

More than 3500 minerals have been identified and only two dozens are very common. These large number of minerals are composed from 92 elements but several factors limit the possible number.

By far most common mineral consists of silicon and oxygen combined with one or more of the other elements listed in following table.

Table 2.2 Common Element in Earth Crust

Element	Symbol	Percentage of Crust by wt.
Oxygen	O	46.6
Silicon	Si	27.7
Aluminium	Al	8.1
Iron	Fe	5.01
Calcium	Ca	3.6
Sodium	Na	2.8
Potassium	K	2.6
Magnesium	Mg	2.1
All other		1.5

Uses of Mineral Resources

Minerals have many uses in everyday life and form a very important part of nation's economy.

Following are the uses of mineral resources:

1. Minerals are used as alloy material in jewellery.
2. Minerals can be used as fungicides, insecticides, seed dressings and as fertilizers.
3. Minerals are used in the formation of alloys, metal plating and coins.
4. Minerals can be used in medicinal system particularly in Ayurvedic system.
5. Minerals can be used in communication system as in the form of cables and electronic devices.
6. Minerals are also used in transportation.
7. Minerals can be used in defense equipment like weapons etc.
8. Minerals are used in development of industrial plants and machinery.
9. Minerals are used for the generation of energy like lignite, uranium and coal.
10. Minerals are used by dentists.
11. Minerals are used for making cooking vessels.

Liquid metal

1. Liquid metals are used in thermometer like mercury.
2. Liquid Metals are used in dental inlays.
3. Liquid metals are also used in electric switches.

Non metals

1. Non metals are used as fertilizers and insecticides. For example, potash.
2. Non metals used in medicines, car battery and industry. For example, sulphur pyrites.
3. Non metals are used in plaster wall board. For example, gypsum.
4. Non metals are used for concrete, building stone, cement industry. For example, lime stone.
5. Non metals are used for construction of sand and gravel, bricks etc. Examples include silicate minerals.

Environmental Effects of Mineral Extraction

Mineral resources extracted from Earth's crust or ocean are not of much importance in relation to environmental effects, rather the impact arises from extraction, processing and use of minerals. Major minerals in India are found in Jharkhand, Orissa, Madhya Pradesh, Andhra Pradesh, Megahalaya, West Bengal, Rajasthan, Karanataka and Tamil Nadu. India produces about 84 minerals.

Six major mines which cause severe environmental damage include:

- East coast bauxite mine in Orissa.
- North east coal field in Assam.
- Kudremukh iron ore mine in Karnataka.
- Sukinda chronite mines in Orissa.
- Joduguda uranium mines in Jharkhand.
- Jharia coal mines in Jharkhand.

Extraction of minerals can be done by two methods

1. sub-surface mining 2. surface mining.

Sub-surface mining involves extraction of minerals from deep deposits in soil. Surface mining involves extraction from shallow deposits. Sub-surface mining is more expensive, destructive and dangerous in relation to environmental impacts.

Important environmental effects of extracting minerals are as follows:

Ground water pollution: Mining by disturbing hydrological cycle contaminates the ground water. Heavy metals get leached down and pollute the ground water. Sulphur from the ores gets leached down into soil and results in acid formation in the ground water.

Surface water pollution: Heavy metals from mining area, acids from mine drainage and radioactive substances from mine wastes contaminate water bodies like rivers, streams, lakes, ponds etc. and kill the aquatic flora and fauna.

Air pollution: During smelting of mineral resources huge amount of air pollutants in the form of suspended particulate matter, sulphur dioxide, arsenic particles, soot, led, cadmium etc. are issued in the atmosphere. These air pollutants cause severe health problems and pollution in the air.

Land subsidence: Land subsidence occurs due to underground mining. It results in cracks in houses, buckling of roads, destruction of rail tracks, leaking of gas pipes etc.

De-vegetation and defacing of landscapes: Deforestation and de-vegetation due to mining leads to environmental losses, in the form of deteoriation of landscapes and soil erosion.

Occupational health hazards: Mine workers suffer from asbestosis, silicosis, fibrosis, black lung diseases, and respiratory diseases due to constant exposure to toxic substances and suspended particulate matter that originate from them.

Table 2.3 Harmful effects of mineral extraction, processing and use.

Steps	Environmental effects
1. Mining extraction	Disturbed land, oil spills, Heat, Destruction of environment, mine waste dumping,
2. Processing Transportation, Purification, and Manufacturing	Noise, Air, Water and Soil pollution, Destruction of environment, mine waste dumping, Heat
3. Use transportation or transmission to individual user, and discarding	Noise, Air, Water, Soil and Thermal Water pollution, Radioactive wastes, Heat and Destruction of environment, mine waste dumping.

Case Study of Udaipur Mine

Mining in Udaipur, spread over 15,000 hectares causes adverse impact on environment. Wash off of minerals, discharge of mine water pollute the nearby areas as well as water of Ahar river. Hills around the mines are devoid of vegetation and cause soil erosion. Blasting activity causes destruction in flora and fauna in the mining area. Discharge of mine water flows towards a big tank known as "Bag Dara". People use this mine water for irrigation purpose due to scarcity of water in the area.

FOOD RESOURCES

There are about 80,000 edible species of plants and only about 30 types of crops feed the world. Four crops, namely, wheat, rice, corn and potato form major part of the world's total food production.

World Food Problems

Producing large quantity of food to feed the world's population is a complex task. There are several interrelated agricultural, economic and environmental food problems. These are as follows:

- Food quantity
- Under nutrition, malnutrition and over nutrition.
- Food storage and distribution
- Poverty
- Food additives
- Environmental effects of agriculture

Food quantity During last 50 years, world food production has increased three times, thereby increasing per capita production by about 50 per cent. At the same time population growth has increased at such a rate that it has outstripped food production. Thus, food problems are directly related to population.

Under nutrition, malnutrition and over nutrition Food and Agriculture Organization (FAO) of United Nations estimated that on an average the minimum calorie intake on a global scale is 2,500 calories/ day. People receiving less than 90 per cent of these minimum dietary calories are called under-nourished. Besides the minimum calorie intake we also need proteins and minerals. People consuming insufficient food often suffer from malnutrition. India is the third largest producer of staple crops and yet an estimated 300 million Indians are still undernourished. Every year 40 million people of under developed countries die of undernourishment and malnutrition.

Food storage and distribution Proper food storage facility, processing, transporting and marketing in most under developed countries is below its potential and much of the food produced never reaches the consumers.

Poverty Poverty does not mean lack of food production. But it is a chief cause of hunger and malnutrition throughout the world. Poor people around the world do not have the requiste land and money to grow enough food for consumption.

Food additives Addition of chemicals to the processed food products are called food additives. These additives reduce spoilage and increase loss of vitamins and other nutrients. For production of food, large quantity of fertilisers, pesticides are used. Very small quantity is used by the crops grown and rest waste pollutes the water ecosystems and depletes the quality of soil. Use of huge quantity of fertilisers enhances burden on the industries hence causing air pollution. Forest land is used for the agricultural purpose and reduces the available biodiversity.

Sometimes food is unevenly distributed within the families, like in India. More food goes to the men while women and children are underfed and malnourished. In urbanised southern Brazil the food supply per person is high while in semi arid north-eastern interior, people are underfed. Hence, two out of three people suffer from malnutrition in this area.

Environmental effects of agriculture More food production has adverse affect on air, water and soil. Severe environmental degradation like soil infertility, deforestation, loss of genetic diversity, water logging, overgrazing etc. takes place.

Case Study of Africa

Africa in 1970 was self sufficient in food but since 1985 due to population growth, flat topograpy, dry climate, less rainfall, overgrazing, deforestation, soil erosion, desertification, shortage of fuel wood, has caused millions of premature deaths each year. If the above said factors are sustained then Africa can be saved from famine, disease, and ecological deterioration.

Changes Caused by Agriculture and Overgrazing

In early days people used to practice of slash and burn cultivation or shifting cultivation. But nowadays, agricultural practices are quite different from the earlier ones and the output, in terms of yield as well as impact on environment is quite different. During early days agriculture involved simple tools, small plot, natural available water for crops and organic fertilizers and resulted in low production. This type of traditional agriculture is still practiced by about half of the global population.

Agriculture Following are the effects of agriculture:

1. **Soil erosion:** Due to deforestation the soil gets exposed to wind, storms and rain which results in loss of top fertile layer.
2. **Deforestation:** Due to shifting cultivation there is burning of trees in forests and results in loss of forest cover.
3. **Loss of nutrient:** Due to the burning of forest, the organic matter in the soil gets destroyed and most of the nutrients are taken up by the crops and makes the soil nutrient poor.

4. **Loss of genetic diversity:** Due to deforestation, loss of genetic diversity takes place because these forests are the home of many wild animals.
5. **Endangered wildlife from loss of habitat:** Due to loss of forest area the wild life gets endangered due to commercial hunting and poaching of animals. The greatest threat to most wild species is destruction, fragmentation and degradation of their habitats. Such disruption of natural communities threaten wild species by destroying migration routes, breeding areas, food resources. Deforestation, especially of tropical forest is the single greatest cause of the decline in global biological diversity by habitat loss.

Overgrazing India possesses a significant position in livestock population in the world. This livestock wealth plays a crucial role in the rural life of our country. Each type of grass land has a herbivore carrying capacity–the maximum numbers of herbivore supported by a given area. Carrying capacity is influenced by season, range, condition annuals, climatic condition, past grazing use, soil type, kinds of grazing animals and duration of animal graze in an area. Light to moderate grazing is necessary for the health of grass land. It maintains water and nutrient cycle needed for healthy grass growth and healthy root system. It hinders soil erosion and encourages build up of organic soil matter. However most often, the grazing pressure is so high that its carrying capacity is crossed and sustainability of the grazed land fails. Following are the effects of overgrazing:

1. **Soil erosion:** By overgrazing the top soil get exposed and eroded by wind and rainfall.
2. **Loss of species:** Overgrazing leads to loss of useful species of plants.
3. **Loss of regeneration capacity:** Due to overgrazing, regeneration capacity of plants is lost. Hence, these species are not regenerated in the ecosystem. The secondary species are less nutritive in nature as compared to primary (original) species. Ultimately the nutritive species are replaced by other thorny species.
4. **Land degradation:** Overgrazing leads to multiple actions resulting in loss of soil structure, hydraulic conductivity and soil fertility. Due to trampling by cattle the soil looses infiltration capacity which reduces percolation of water into the soil.

Effects of Modern Agriculture

Modern agriculture involves the use of hybrid seeds, high-tech equipments, fertilizers, pesticides, and irrigation water. Modern agriculture has succeeded significantly in increasing crop yield (green revolution).

Green Revolution: The introduction of scientifically bred or selected varieties of grain (rice, wheat, maize with high enough inputs of fertilisers, and water can greatly increase crop yield). Green revolution was possible because of:

1. **Use of high yielding varieties:** In modern agriculture the use of high yielding varieties encourages monoculture. If a disease attacks, there is total devastation of the crop due to monoculture practice.
2. **Fertilizer and pesticides related problems:** Fertilizers used in modern agriculture include phosphorus, potassium, nitrogen etc. Farmers use these fertilizers to increase the growth of crop. Excessive use of fertilizers may cause micronutrient imbalance. Following are the problems related to the use of fertilizers:
   (i) **Pollution** Nitrate used in the fields gets leached down into deep soils and contaminate the ground water. The excessive levels of nitrate ions makes drinking water toxic, especially for infants. The disease is known as **baby syndrome** or **metahaemoglobinemia**. This disease causes death of infants.
   - **Water pollution:** Due to excessive use of NPK, run off from the fields to nearby water bodies causes nourishment of the lakes. Due to eutrophication, lakes get invaded by algal blooms. This causes death and decay of flora and fauna in lakes.
   - **Reduced oxygen content of soil:** Excessive use of NPK fertilizer also reduces oxygen content of soil and alters its porosity.
   - **Reduced natural nitrogen production efficiency:** Excessive use of NPK fertilizers causes the soil to become compact and less suitable for crop growth and reduces its natural ability to produce nitrogen in forms usable by plants.

- **Reduced humus production:** Excessive use of organic and inorganic fertilizers also reduces humus production in the soil.

(ii) **Problems related to Pesticides** Various chemicals used to control pest population are called pesticides. Paul Mueller (1939) discovered insecticidal properties of DDT. There are thousands of pesticides used in agriculture. The first generation pesticides include chemicals like arsenic, lead, mercury and sulphur to kill the pests while second generation pesticides are DDT. Although pesticides are protecting our crop yet they have created a number of problems which are discussed as:

- **Biological magnification:** Increase in concentration of pesticides and other slowly degradable fat soluble chemicals in organisms at successively higher trophic levels of food chain or food web.

  Hence the level of pesticides in human being is much higher than the primary trophic level. Pesticides are non-biodegradable and they keep on accumulating in the food chain. This is called **biological magnification**. The biomagnified form of pesticides in the human beings is very harmful. Since human beings occupy the higher tophic level in the food chain.Use of pesticide causes destruction of wildlife. Accumulation of these pesticides in fatty tissue of the organism causes mutation in the reproductive cells. Thining of egg shell by excessive use of pesticides results in egg hatch before time causing death. Pesticides are carcinogenic and responsible for destruction of wildlife.
- **Threat to wildlife:** Wildlife gets destroyed due to the use of pesticides.
- **Development of genetic resistance:** Some individuals of the pest species usually survive even after the pesticide spray and give rise to development of genetic resistant pest species.
- **Production of new pest:** About twenty new species of pests are known which became resistant to all types of pesticides and are called as "superpest"

- **Death to non-target species:** Many insecticides have broad spectrum poisons which not only kill the target species but also the non-target useful species.
- **Threats to human health:** Due to excessive use of pesticides in contaminated food, human health is threatened.

Water Logging

Water logging is the saturation of soil with irrigation water so that the water table rises close to the surface. World wide, about one-tenth of all irrigated land suffers from water logging. Under waterlogging conditions, porous spaces in the soil are drenched with water and soil air gets depleted. The water table rises and roots of plants do not get adequate air for respiration. Crop plants get lodged, mechanical strength of the soil declines and results in low crop yield. Water logging can be controlled by preventing excessive irrigation, sub-surface drainage technology and bio-drainage with trees like Euclyptus.

Salinity Problem

Salinity is the quantity of salt dissolved in a given volume of water. Accumulation of salts in soil can eventually make the soil incapable of supporting plant growth. This is called **salinisation**. One-third of the world's irrigated land is now affected by salinity. In India, about seven

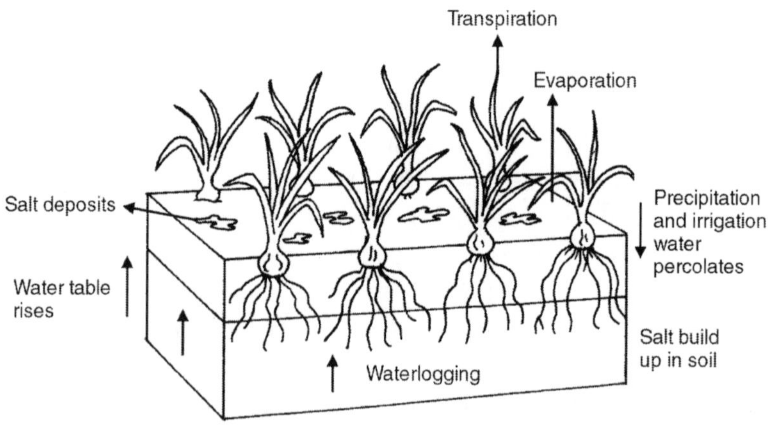

Fig. 2.3 Diagram showing waterlogging and salinisation

million hectares of land is affected by salinity. Saline soils are characterised by having electrical conductivity more than 4 ds/m and pH exceeds 8.0 and the exchangeable sodium percentage is more than 15 per cent. Major cause of salinisation is excessive irrigation. About 20 per cent of the world cropland receives irrigation with canal and ground water which is usually contaminated by dissolved salts. During dry climate, water evaporates leaving behind salts in the top soil profile.

Salinisation cause:

- Decreased crop yield.
- Stunted crop growth and at times kills them.
- Land becomes unproductive.
- Hazardous to wild life, especially water fowl.

ENERGY RESOURCES

Energy resources are renewable as well as non-renewable resources. Energy is considered to be an index of development for a nation as the developmental activities of the nation depends upon the consumption of energy in that particular country.

Growing Energy Needs

Development of different countries depends upon input of energy. Industry, mining, agriculture, lighting, cooling and heating in buildings, household, transportation all need energy. The fossil fuels like oil, petroleum, coal and natural gas provide 95 per cent of the energy. Our life style has changed from a simple one to a luxurious way of living. It leads to increased consumption of energy in our homes. Our standard of living are closely related to energy consumption. Norway, Switzerland and U.S.A. have high gross national products and energy consumption while India and China have low gross national product and low energy use. Bahrain and Quatar are oil rich states and hence their energy consumption and GNP are more but their development is of those standards.

Renewable and Non-renewable Energy Sources

A source is one that provides adequate amount of energy from the environment to meet human needs and wants over a longer period of time.

The sources are of two kinds:

1. **Renewable resources:** Resources that can be generated from the natural processes are called renewable resources. Example are trees in forests, grasses in grasslands, fresh water in lakes, deposits of ground water, fresh air, and fertile soil. These are inexhaustible in nature. Solar energy, tidal energy, hydropower, biomass, geothermal energy and hydrogen. These are also called as non-conventional sources of energy because they can be used again and again.
2. **Non-renewable resources:** Resources that are available in a fixed quantity in the Earth's crust are called non-renewable resources. They are exhaustible in nature because they are not replaced by natural processes. Examples are coal, petroleum, natural gas, nuclear fuels like uranium. These are also called as conventional sources of energy because they cannot be used again and again in an endless manner.

Renewable Energy Sources

Use of alternate energy sources Important alternate energy resources are described below:

1. **Solar energy:** Sun is renewable and inexhaustible source of energy. It is a pollution free source. Human beings use solar energy in day to day life in various forms like drying of clothes, production of food grains, preservation of eatables, for obtaining salt from sea water, in the form of light, fan, geyser and cooker etc. There are many methods by which we can extract solar energy and convert it into electrical or mechanical energy.

   These are discussed as:

   Solar cells: Solar cells convert solar energy into electrical energy. These solar cells are made up of silicon and gallinium. When solar radiations fall on the solar panels they get converted into electric energy. Solar panels are made up of silicon cells.

   Solar heat collectors: Solar heat collectors are made up of stones, bricks etc. These solar heat collectors absorb heat during the day time and release it slowly at night.

Tidal energy: Tidal waves are formed by gravitational forces of sun and moon. The height of the tides are used for spinning the turbines, which generate electricity.

Wind energy: Wind energy is produced by windmill. But windmills are restricted to certain coastal areas, open grassland or hilly region, Windmills are used for grinding grains and pump water.

Hydropower energy: Hydropower energy is also known as hydroelectric energy. Hydroelectric energy is produced from the kinetic energy of water falling from a certain height. This type of energy is produced in big dams. Water falling from a height rotates the turbines at the bottom of the dam to generate electricity. It is pollution free energy. Hydropower potential of India is about 4 × 10" KWH.

Geothermal energy: This energy is generated from the hot rocks present inside the Earth. For extracting geothermal energy, a hole is drilled in hot rocks and by putting pipes inside, the steam or hot water is extracted through the pipe at high pressure which rotates the turbine to generate electricity.

Ocean thermal energy: It is also denoted by OTE. In this, energy is generated due to differences in temperature of water at the surface of the ocean and at deeper levels. The warm surface water boils the liquid ammonia and propane. The high pressure vapor of the liquid formed by boiling are used to rotate the turbine and produce electricity.

2. **Biomass energy:** Biomass is the term used for all materials produced from photosynthesis. Biomass involves residues, plants, wastes, marine algae, aquatic plants, and forest residue etc. Biomass energy constitutes renewable resources and inexhaustible form of energy. It is a sink for carbon dioxide which help to conserve soil and water.

   Types of Biomass energy are discussed as:

   Biogas: It is a pollution free and clean technology. In this a lot of animal waste and agricultural waste is put for anaerobic degradation in the presence of water. Anaerobic degradation is the breakdown of organic matter by bacteria in the absence of oxygen. This anaerobic degradation results in the production of biogas. Biogas is the mixture of methane, carbon dioxide, hydrogen and hydrogen suphide. These gases can be used as fuel.

Petro-plants: The hydrocarbon rich plants are converted into petroleum hydrocarbon under high temperature and pressure. The products obtained from these petro-plants are gases, kerosene, oil cake etc. Examples are Euphorbia plants which are rich in hydrocarbons.

Energy plantation: It is also known as **dendrothermal energy**. In this wastelands are used for plantation of fast growing trees, shrubs and herbs with high calorific value. These plants produce energy either by burning or may be converted into fuels by fermentation.

Energy from urban and agricultural waste: Wastes of sewage and agriculture are converted into gas and electricity. Agricultural wastes are burnt in dweller for generating fuel.

Bagasse based plants: Bagasse, a waste of sugar mills can be used to generate energy. Biomass can be fermented to alcohol like ethanol and methanol which can be used as fuel.

Non-renewable energy resources These types of resources involve fossil fuels like natural gas, petroleum, coal and nuclear fuels.

1. **Natural gas:** This is a fossil fuel. It is produced by decomposed remains of dead animals and plants buried under the earth. The main constituents of natural gas are methane, propane and ethane. Natural gas is easily transported and pollution free fuel. It is used as a domestic and industrial fuel.
2. **Petroleum:** It is a cleaner fuel and pollution free and is easy to transport. Petroleum is a mixture of alkane hydrocarbons. After purification by fractionation and distillation at different temperatures, it gives rise to a number of products like petroleum gas, kerosene, petrol, diesel, fuel oil, lubricating oil, paraffin wax, asphalt, plastic, and so on.
3. **Coal:** Coal is an abundant fossil fuel in world. It is produced from the buried dead plants and animal remains. These remains undergo high heat and pressure and convert into peat and coal over millions of years. India has about 5 per cent of the world's coal.

4. **Nuclear energy:** Nuclear energy is a fossil fuel and has tremendous potential. Two methods, namely, nuclear fusion and nuclear fission generate nuclear energy.

LAND RESOURCES

Land forms about one-fifth of the earth surface. Land is a valuable and unlimited resource. Land is largely covered with grasslands, wetlands, natural forests, agricultural land, urban and rural settlements. The top layer of the land is soil and continuously regenerated by natural process over a period of time. Hence classified as a renewable resource. Soil is a most important resource. Plants and animal materials decay and are released into the nutrient bank of the soil. Soil development process takes several decades to complete. Two process, namely, weathering and pedogenesis form soil.

Land Related Problems

Land degradation Human population is expanding day by day and its need for more land to produce fiber, fuel wood and food is also increasing at an alarming rate. Land degradation is of great concern because soil reformation is a very slow process. The annual erosion rates for agricultural land is about 18 to 100 times more than the renewal rate. Tracts of land have been cleared of the natural vegetation. Lack of natural vegetation results in soil erosion. Land degradation also depends upon topography, wind intensity, rainfall and the type of agricultural practices. There are so many causes of land degradation like water logging, deforestation, salinisation, contamination of soil with sewage and industrial effluent.

Man induced landslides Man induced landslides usually take place in hilly areas. While in plains, landslides are due to construction of large buildings, mining, reservoirs, large dams, construction of roads and railway tracks are the major causes of land slides over hilly as well as plain areas. Examples of man induced landslides are in the areas between Rishikesh and Byasi on Badrinath highway in India. During landslides large portions of fragile mountains are destroyed and they fall into adjacent valleys and streams. Due to falling of landmasses over the adjacent mountains and valleys, fragility of the land and turbidity of the nearby streams increases.

Soil erosion Soil erosion is the movement of soil components from one place to another. Soil erosion results in loss of fertility of topsoil layer. About 7 per cent of the world top soil washes and blows into the rivers, lakes and oceans each year. So a large amount of the potential of cropland is eroded. The plant nutrient loss from the erosion is about $18 billion per year.

Types of Soil Erosion:

Soil erosion is mainly of two types:

1. **Natural geologic erosion:** It is removal of topsoil by natural process. It maintains the balance between erosion and renewal.
2. **Human-accelerated erosion:** It occurs due to human activities like overgrazing, mining and deforestation. In this, rate of erosion is faster than the renewal.

Causes of soil erosion

There are mainly two types of agents which cause soil erosion:

1. **Biotic agents:** Biotic agents are overgrazing, deforestation and mining. About one-fourth of the world's soil is eroded by the biotic agents.
2. **Climatic agents:** Wind and water are two climate agents responsible for soil erosion.

   Water erosion: Water erosion is the movement of soil by rainwater, melted snow etc. It is of many types as discussed below:

   (i) **Rill erosion:** When rainwater produces rills or finger shaped groves over the area.

   (ii) **Sheet erosion:** It involves more or less uniform removal of this layer of soil from the area.

   (iii) **Gully erosion:** When rainwater forms deeper cavities or gullies.

   (iv) **Stream bank erosion:** During rainy season, water of streams, lakes turn in other direction by cutting the soil.

   Wind Erosion: Wind erosion is common in barren lands. It takes place by the soil movements along with the high velocity wind. These are of three types namely -

   (i) **Saltation:** In this 1-1.5 mm diameter of soil particles or in the from of short bounces moves along with the wind.

(ii) **Suspension:** In this less than 1 mm diameter of tiny soil particles move along with the high velocity of wind.

(iii) **Surface creep:** In this 5 to 10 mm diameter heavy soils particles move along with the wind.

Desertification It includes a number of ecological changes that destroy the ability of land to be used for growing crops or grazing. It is the human activities that degrade soil and water resources and lead to the creation of desert like land. The conversion of rangeland, rain fed cropland, or irrigated cropland to desert like land with a drop in ten per cent agricultural productivity is called **desertification**.

There are three types of desertification:

1. **Moderate desertification:** It reduces productivity up to 10% to 25%.
2. **Severe desertification:** It reduces productivity up to 25% to 50%.
3. **Very severe desertification:** It reduces productivity more than 50% and results in formation of sand dunes.

Causes of desertification

- drought and hot temperature.
- overgrazing.
- deforestation.
- improper soil and water resource management.
- salinisation.
- water logging.
- mining.

Most affected areas of desertification are South Western United States, Central and South America, Africa, Asia and Australia. According to UN environmental program one-fifth of the world land area may become desertified over the next 20 years.

ROLE OF AN INDIVIDUAL IN CONSERVATION OF NATURAL RESOURCES

Conservation of natural resource involves wise use and careful management of resources, so as to obtain the maximum possible social benefits from them for present and future generations. Conservation makes important contributions to social and economic development. Methods used for conservation includes preservation, balanced multiple use, reducing unnecessary waste, recycling, reuse and decreased resource use. Conservation efforts are underway at national as well as international level. Environment belongs to all of us and each one of us have a responsibility to contribute towards its conservation. So individual efforts together help in conserving our natural resources to some extent. Following are the methods by which we can help conserve natural resources.

Methods of Conservation of Resources

Conserve energy

- Construct house with more doors and windows to provide light.
- Use solar cooker, solar light and solar geysers.
- Turn off lights, fans and other electrical appliances after use.
- Grow big trees for cooling effect in summers.
- Recycle and reuse metals, paper and glass.
- Use bicycle for riding instead of car and scooter.

Conserve soil

- Develop a garden in open areas to prevent soil erosion.
- Grow trees in colony to prevent erosion.
- Make compost of kitchen waste.
- Use mixed cropping for the protection of soil nutrients.
- Use cow dung for the manuring of the fields.
- Reduce the use of pesticides, fungicides.
- Use drop irrigation for watering.

Conserve water

- Use washing machines for washing clothes.
- Install water saving toilets.
- Don't keep water taps running while not in use.
- Check for water leakage in pipes and toilets.
- Reuse the soapy water for courtyards.
- Install a water harvesting system in your house.
- Use drip irrigation.

Equitable Use of Resources for Sustainable Development

There is big gap between the low developed countries (LDC) and more developed countries (MDC) in terms of population growth, industrial growth, use of natural resources, energy, pollution, etc. Sustainable development means use of resources in judicious manner so that future generation is able to use the resources.

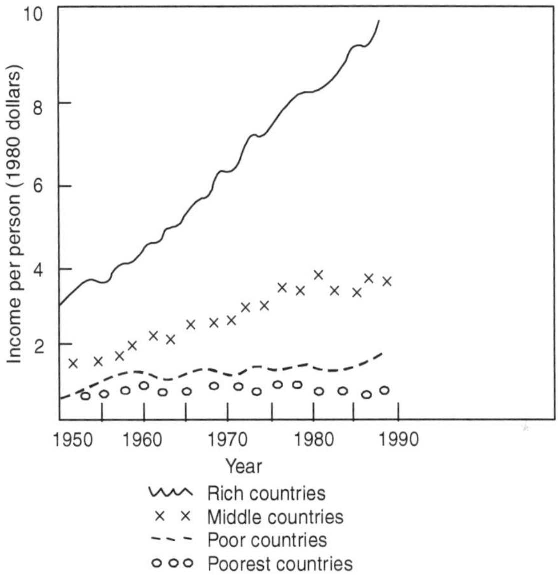

Fig. 2.4 Wealth gap (Data from United Nations)

Less developed countries are culturally very rich but economically they are less developed as compared to more developed countries. As more developed countries are developing further, leading to more pollution and sustainability of the earth, life support system is under threat. The low developed countries have more population and are struggling hard with poverty problems. The depletion of resources is leading to un-sustainability of the country.

The main reason for un-sustainability are the over population and under consumption of natural resources by poor countries. So, in order to attain sustainable development there should be equitable use of resources. So more developed countries should lower down the consumption of resources and generation of wastes. This will help in building a better world.

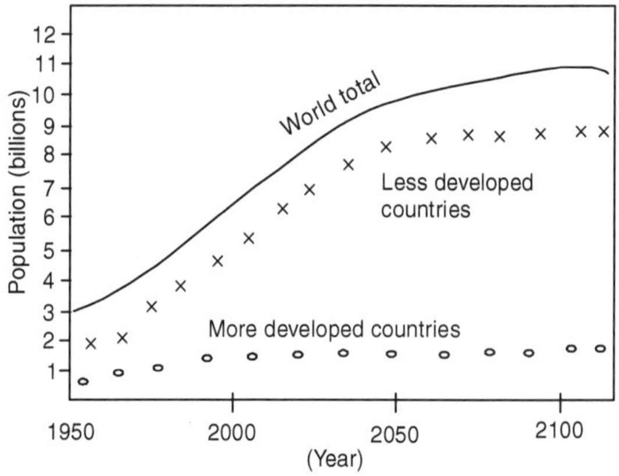

Fig. 2.5 Population size for MDCs and LCDs and the World (Data from United Nations)

REVIEW QUESTIONS

Short Type Questions

1. What are the natural resources?
2. How many types of natural resources are there?
3. Define deforestation.
4. What is Jhum cultivation?
5. What is timber extraction?
6. What are the effects of mining?
7. What is mining?
8. Define:
   (a) Water resources
   (b) Surface water
   (c) Aquifers
   (d) Flood
   (e) Drought
   (f) Renewable and non-renewable resources
   (g) Desertification
9. What is ground water?
10. Give various types of acquifers.
11. What are mineral resources?
12. Give various types of minerals.
13. What are food resources?
14. What is water logging?

15. What are energy resources?
16. What is source?
17. What do you mean by Biomass energy?
18. What is energy plantation?
19. What are petro-plants?
20. What is nuclear energy?
21. What are land resources?
22. What do you mean by soil erosion?
23. Name different type of soil erosion.

Descriptive Questions

30. What are the forest resources? Explain its uses in detail.
31. Write the importance of forests.
32. What is the difference between desertification and deforestation?
33. What are the causes of deforestation?
34. Explain the consequences of deforestation.
35. Discuss the effects on forests and the tribal people.
36. Explain the effects of Dams on forest and tribal people.
37. How water can be prevented from overexploitation?
38. What are uses of water?
39. Describe the preventive methods of flooding.
40. Discuss the conflicts over water.
41. Explain the benefits of dams.
42. What are the environmental problems of constructing big dam?
43. Discuss in detail the uses and exploitation of mineral resources.
44. What are the environmental problems that occur during extraction of minerals?
45. Explain the uses of minerals.
46. Discuss the environmental effects of producing more food resources.
47. What is overgrazing and discuss its effects?
48. What are the effects of modern agriculture?
49. Explain in detail the problems of salinity.
50. What are the growing energy needs of world?
51. Discuss in brief the alternative energy sources.
52. Write short notes on the following
    (a) Non renewable energy resources.
    (b) Land degradation.
    (c) Man induced land slides.
53. Discuss in detail the causes of desertification.
54. What is the role of an individual in conservation of natural resources?

3

Ecosystems

CHAPTER OBJECTIVES

- Concept of Ecosystem
- Features of Ecosystem
- Ecosystem Functions
- Ecosystem in Various Habitats

German biologist Ernst Haeckel in 1869 coined the term *ecology*. This term is made up of two words *Oikos* meaning house and *logos* meaning study of. Hence, ecology is the study of living things in their environment or it is the study of interactions between living organisms and their environment. Hence, ecology has also been called as *environmental biology* by Snutti 1977.

CONCEPT OF ECOSYSTEM

The term Ecosystem was first coined by A. G. Tansley in 1935. He defined ecosystem as the system resulting from the integration of all the living and non-living factors of the environment. The literal meaning of *eco* refers to the environment, and *system* refers to an interacting and interdependent complex.

The organisms of any community, besides interacting among themselves, always have a functional relationship with the environment. This structural and functional system of communities and environment is called *ecological system or ecosystem*. It is the basic functional unit in ecology, since it includes both biotic and abiotic environment, influencing each other for the maintenance of life. According to George Usher 1965, the complete ecological system of an area, including the plants, animals and the environmental factors is known as ecosystem.

According to Odum 1971, any unit that includes all the organisms (*i.e.* the community) in a given area interacting with the physical environment, so that a flow of energy leads to clearly defined trophic structure, biotic diversity and material cycles (*i.e.* exchange of materials between living and non living components) within the system, is called an ecosystem.

According to a Dictionary of the Environment written by Michael Allaby 1983, a community of interdependent organisms together with the environment is known as ecosystem. Ecosystem includes interacting and interdependent components that are open and linked to each other. In a simplest ways all ecosystem can be defined as a self- sustained community of plants and animals existing in its own environment. An ecosystem may be as small as a drop of pond water or as large as an ocean.

Earth is a good example of a giant ecosystem where both abiotic and biotic components are constantly acting and reacting upon each other. For convenience, Earth as an ecosystem is studied under various smaller

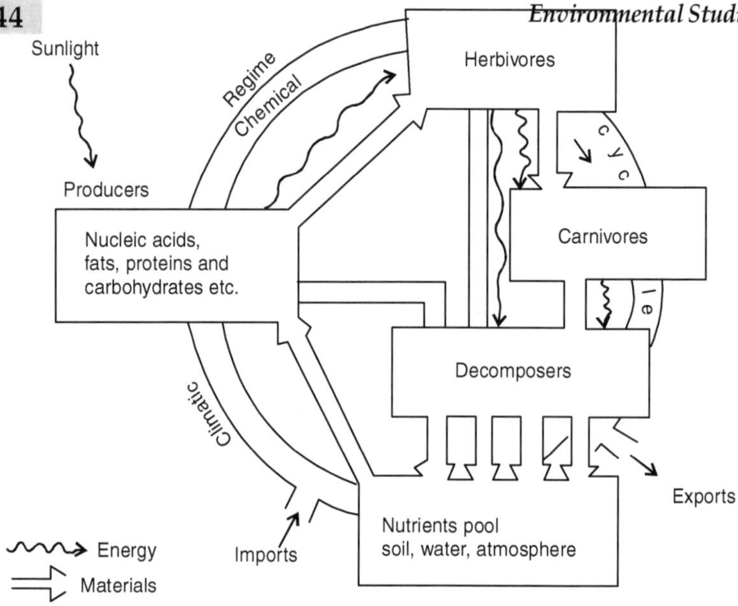

Fig.3.1 Model of ecosystem showing structure and function

ecosystems, for example, grass land, crop land, desert, forest, fresh water and marine water etc.

Thus, ecosystem is an integration of mosaic of interacting organisms and their environment. It is an open system with a continuous, but variable influx and loss of materials and energy. An ecosystem represents the highest level of ecological integration, which are energy based and capable of energy transformation, accumulation and circulation. Its main function in ecological sense is to emphasise obligatory, interdependence and casual relationship, that is, coupling of components to form functional units. So, ecosystem may also be regarded as energy processing unit which is restrained or limited by the account of nutrients and water available to it.

FEATURES OF ECOSYSTEM

The following are the main features of ecosystem according to Smith (1966):

- Ecosystem structure is related to diversity of species.
- It is a major structural and functional unit of ecology.
- Ecosystem function is related to energy flow and material cycling through and within the system.

- The relative amount of energy needed to maintain an ecosystem depends on its structure. The more complex the structure, the lesser the energy it needs to maintain itself.

Types of Ecosystem

There are two types of ecosystem:

Natural Ecosystem
Artificial Ecosystem

Natural ecosystem These operate under natural conditions and are of two types.

1. **Terrestrial ecosystem** It includes forest, desert, grassland system.
2. **Aquatic ecosystem** These are of two types.

   Fresh water: Fresh water ecosystem are of three types:
   Lentic: These are standing water like lakes and ponds as shown in Figure 3.3.
   Lotic: These are flowing water and includes rivers, streams etc.
   Wetlands: These incudes marshes, swamps and bogs.
   Marine water: These are of of three types:
   Open oceans: It includes pelagic
   Upwelling regions
   Coastal areas: It includes wetlands, estuaries, coral reefs and mangroves.

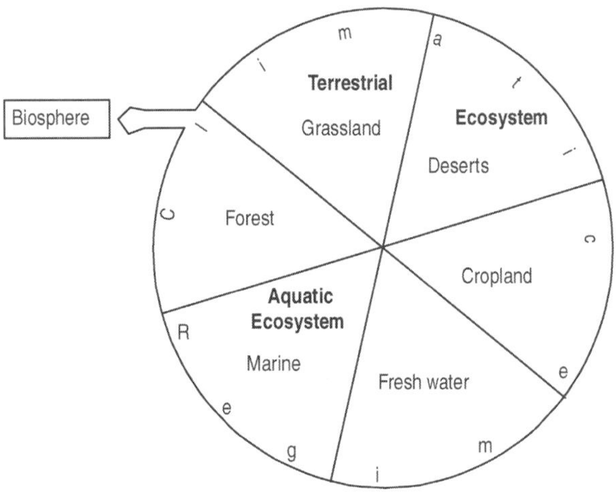

Fig.3.2 Diagram showing terrestial and aquatic ecosystem

Structural Components of Ecosystem

Structural ecosystem consists of two major components:

Abiotic components It consist of non-living substances such as inorganic phosphorus, sulphur, carbon, nitrogen and hydrogen etc. Abiotic components also include physical and chemical factors.

Physical factors of the ecosystem are as follows:

1. Average temperature
2. Sunlight and shade
3. Precipitation
4. Wind
5. Altitude
6. Latitude
7. Fire
8. Soil
9. Suspended solid material in aquatic ecosystem
10. Water currents

Chemical factors of ecosystem include:

1. Level of plant nutrients in terrestrial and aquatic ecosystem.
2. Level of water and air in the soil.
3. Salinity of water.
4. Amount of dissolved oxygen in aquatic ecosystem.
5. Level of toxic substances in ecosystem.

Biotic components Organisms and green plants consist of biotic component. The organisms are distinguished on the basis of their nutritional relationships. These are classified as follows:

Autotrophic component: These are also called as producers. This component includes producers or green plants only. These are self-feeders. Self-feeders are the organisms that can manufacture their food with the help of sun energy and the process is known as **photosynthesis**.

$$\text{Carbon dioxide} + \text{Water} \rightarrow \text{Glucose} + \text{Oxygen}$$

In this process, the energy of the Sun is converted into chemical energy in the form of carbohydrates. This stored energy is utilised by organisms in the form of food. About 59 per cent of photosynthesis takes place on land while 41 per cent is in the oceans and fresh water systems.

Some organisms such as bacteria convert inorganic compounds into organic compounds without the presence of sunlight. This process is known as **chemosynthesis**.

Heterotrophic component: This component includes consumers or heterotrophs. These organisms consume the matter synthesized by the producer.

Some consumers feed on living plants and animals and some feed on dead plants and animals which leads to the formation of different categories.

Based on their nutritional sources, consumers are divided into following categories:

1. **Herbivores:** They are plant eaters and are known as primary consumers. These organisms feed on living plants. Such as rabbits feed on leaves, birds feed upon seeds. There are some Zeeplanktons that feed on phytoplanktons.
2. **Carnivores:** These are the flesh eaters and are known as secondary consumers. These organisms feed on plant eating animals or herbivores such as birds and spiders which in turn feed on plant-eating insects. Tertiary consumers are those that feed on animal-eating animals such as hawks feed upon snakes and sharks, which feed on other fishes.
3. **Omnivores:** These are known as **everything eaters**. These feed upon both plants and animals. Examples are rats, cockroaches, pigs etc.
4. **Detritivores:** These are consumer organisms that feed on detritus and are known as detritivores. These are divided into two classes: decomposers and detritus feeders.

FUNCTIONS OF ECOSYSTEM

Function of ecosystem may be classified as:

- Food chain
- Ecological pyramid
- Energy flow
- Ecological succession

Food Chain

The transfer of food energy from the producers, through a series of organisms, with repeated eating and being eaten is called **food chain**. Thus green plants occupy, in any food chain, the first trophic level–the producer level, and are called the primary producers. The energy, as stored in food matter manufactured by green plants, is then utilised by the plant eaters–the herbivores, which constitute the second trophic level–the primary consumer level, and are called primary consumers or herbivores. Herbivores in turn are eaten by the carnivores, which constitute the third trophic level–the secondary consumers level and are called the secondary consumers or carnivores. These inturn may be eaten still by other carnivores at tertiary consumer level, that is, by the tertiary consumer or carnivores. Some organisms are omnivores that eat the plants as well as animals at their lower level in the food chain and they may occupy more than one trophic level in the food chain. Thus in any food chain, energy flows from producers → primary consumers → secondary consumers → tertiary consumers and so on.

The efficiency of energy transfer also varies from one trophic level to another. At every step large amounts of the potential energy is lost through the dissipation of heat.

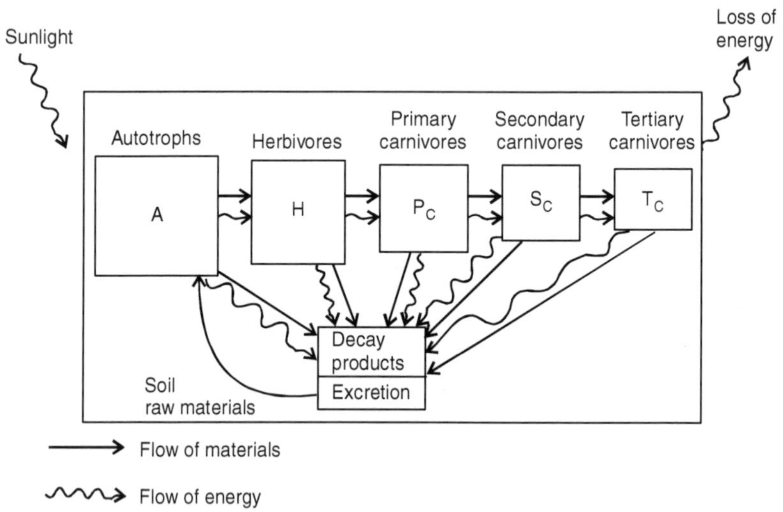

Fig. 3.3 Diagram showing nutritional relationships with different tyrophic levels

In nature, generally there are three types of food chains:

1. **Grazing food chain:** This type of food chain starts from living green plants, and ends at detrivores through grazing herbivores. This food chain is dependent on influx of solar radiation. This food chain is common in terrestrial and aquatic ecosystems where most of the primary production is edible by herbivores.

Table 3.1 Examples of Grazing Food Chain

Type of ecosystem	Producer consumer	Primary consumer	Secondary consumer	Tertiary
Aquatic	Phytoplankton→	Zooplankton→	Fish →	Fish (ban)
Aquatic	Phytoplankton→	Fish →	Frog →	Raccoon
Terrestrial				
a)Terrestrial	Grass →	Grass hopper →	Frog →	Snake
b) Terrestrial	Grass →	Grass hopper →	Birds →	Hawk
c) Terrestrial	Grass →	Rabbit →	Lion →	-
d) Terrestrial	Grass →	Pig →	Man →	-

2. **Parasitic food chain:** It also starts from green plants, then goes to herbivores which may be the host of a large number of lice which live as ecto-parasites.
3. **Detritus food chain:** The food chain moves from dead organic matter to micro-organisms and then to the detritus feeding organisms and their predators. This type of food chain is known as detritus food chain. They are dependent on the influx of organic matter produced in another ecosystem.
   Dead organic matter → Fungi → Squirrel → Hawk
   Dead organic matter → Earthworm → Frog → Snake → Bird

Food Web Food chain normally does not operate in isolation but are interlocked with each other forming some sort of pattern known as Food Web. An organism in the ecosystem may operate at more than one trophic level, that is, it derives its food from more than one source and in turn, may serve as a source of food for several organisms of higher trophic level. This results into two or more chains intersecting each other. The second reason for the formation of food web seems to be the successive loss of energy at higher trophic levels till no more energy is available to support yet another link in the food chain. Food web is important in maintaining the stability of an ecosystem in nature.

Food web in a grass land ecosystem

	Grasshopper	→	Lizard	
Grass →	Rabbit	→	Hawk	
	Mouse	→	Snake	

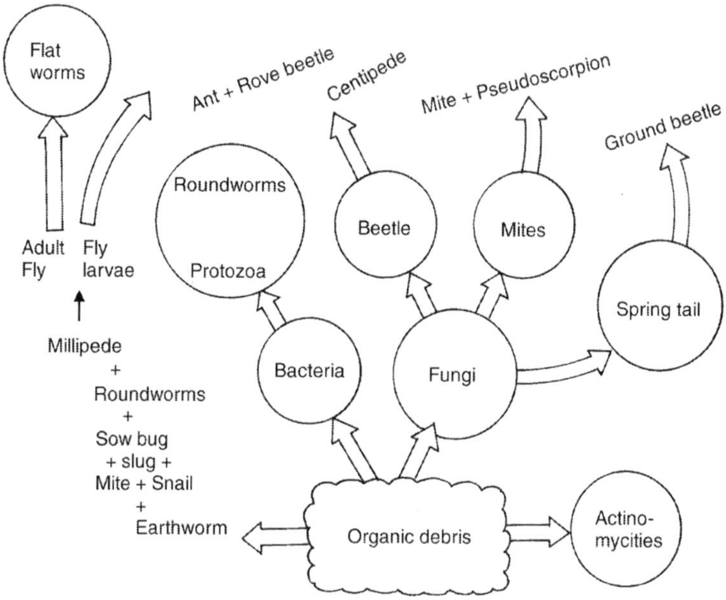

Fig.3.4 Food web of living organisms in soil

Ecological Pyramids

This concept was given by Charles Elton (1927). There is some sort of relationship between the number, biomass, and energy content of the primary producers, consumers of the first and second orders and so on to the top carnivores in the ecosystem. This relationship may be represented graphically by means of pyramids which have been referred to as **ecological pyramids**, where the producer level forms the base of the pyramid and the successive levels makes the apex.

Types of Ecological Pyramids

1. **Pyramid of numbers:** It shows the number of organisms at each trophic level. The relationship between the number of producers,

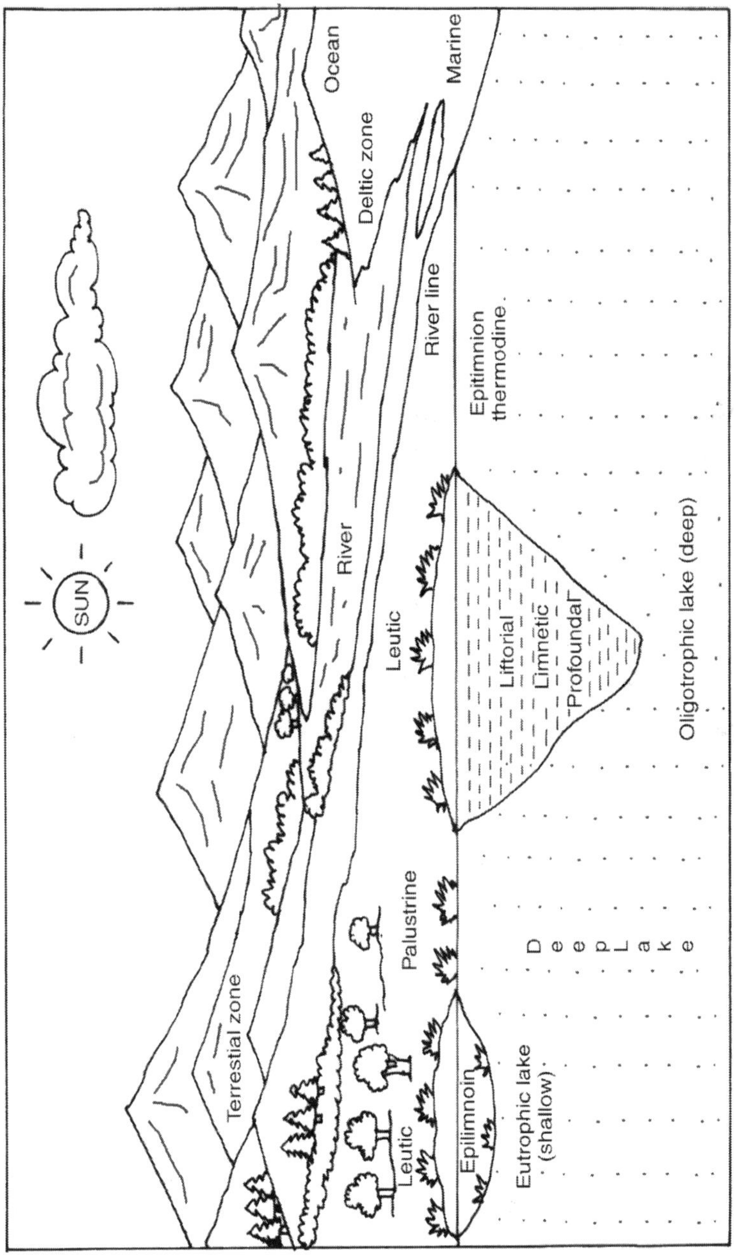

Fig.3.5 Upright pyramid of numbers of (a) Grassland ecosystem and (b) Crop ecosystem

consumers of primary, secondary and tertiary orders constitutes the pyramid of numbers.

In a grassland, the producers are mainly grasses, always maximum in number. This number shows a successive decrease towards apex, as the primary consumers, which are rabbits, mice, etc. are lesser in number than the grasses, the secondary consumers, the snakes and lizards are lesser in number than the rabbits and mice. Finally, the top consumers, the hawks and birds, are the least in number. Thus, the pyramid becomes upright. The pyramid is upright in pond ecosystem also. But in a forest ecosystem the pyramid of number is of different shape because the producers which are mainly large sized trees are lesser in number, and form the base of the pyramid. The herbivore, which are the fruit eating birds, deer etc. are more in number than the producers. Then, there is a gradual decrease in the number of successive carnivores, thus making the pyramid upright.

In a parasitic food chain, the pyramid is always inverted. This is because a single plant supports the growth of many herbivores and each one of these, in turn, may provide nutrition to several hyper-parasites like bugs and lice. Thus from the producers towards consumers, the number of organisms successively show an increase, making the inverted pyramid as shown in Fig. 3.6.

2. **Pyramid of biomass:** It depicts the total dry weight. It represents the relationship between different trophic levels in terms of weight of organisms. In grassland and forest, there is a gradual decrease in mass of organisms at successive levels from the producers to the top consumers. Thus the pyramid is upright. In aquatic ecosystem the biomass of producer is least. This value gradually shows an increase towards the apex of the pyramid, thus making the inverted pyramid as shown in Fig. 3.8.
3. **Pyramid of energy:** It shows the amount of energy flow at successive trophic level. The pyramid of energy represents the total quality of energy utilized by different trophic level organisms of an ecosystem per unit area over a set period of time. The shape of the pyramid of energy is always upright because in this the time factor is taken into account as shown in Fig. 3.9.

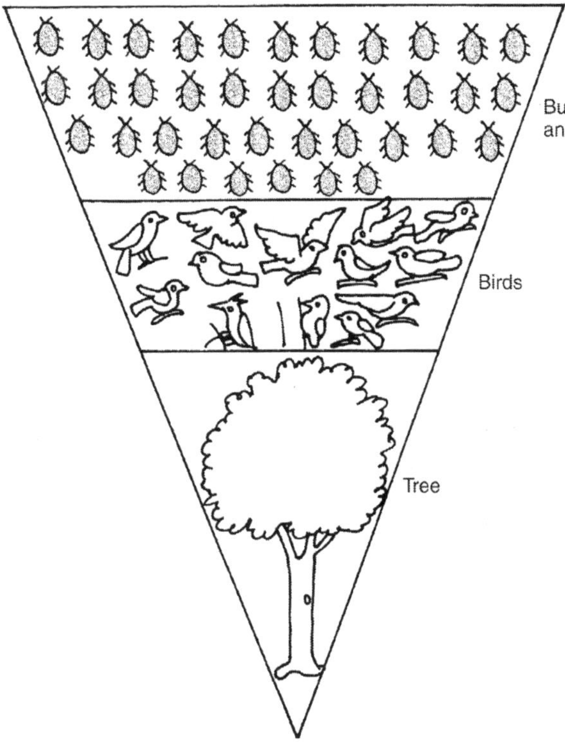

Fig. 3.6 Inverted pyramid of number of a Tree ecosystem

Energy Flow

After having an idea of producers and consumers and their linking in the food chain, it is important to know how energy flows from one trophic level to another trophic level. For proper energy flow from one level to another trophic level, firstly, there should be utilization of energy from sunlight. Secondly, there should be procurement of inorganic substances from environment. In both of these processes energy is never restored. Energy is transfered from one trophic level to another trophic level. Only small amount of the energy is changed from one form to another and large amount of energy is wasted in the form of respiration and other processes. Actually energy flow follows the first law of thermodynamics. According to this law, energy can neither be created nor destroyed, rather, it transfers from one form to another form. The transfer of energy

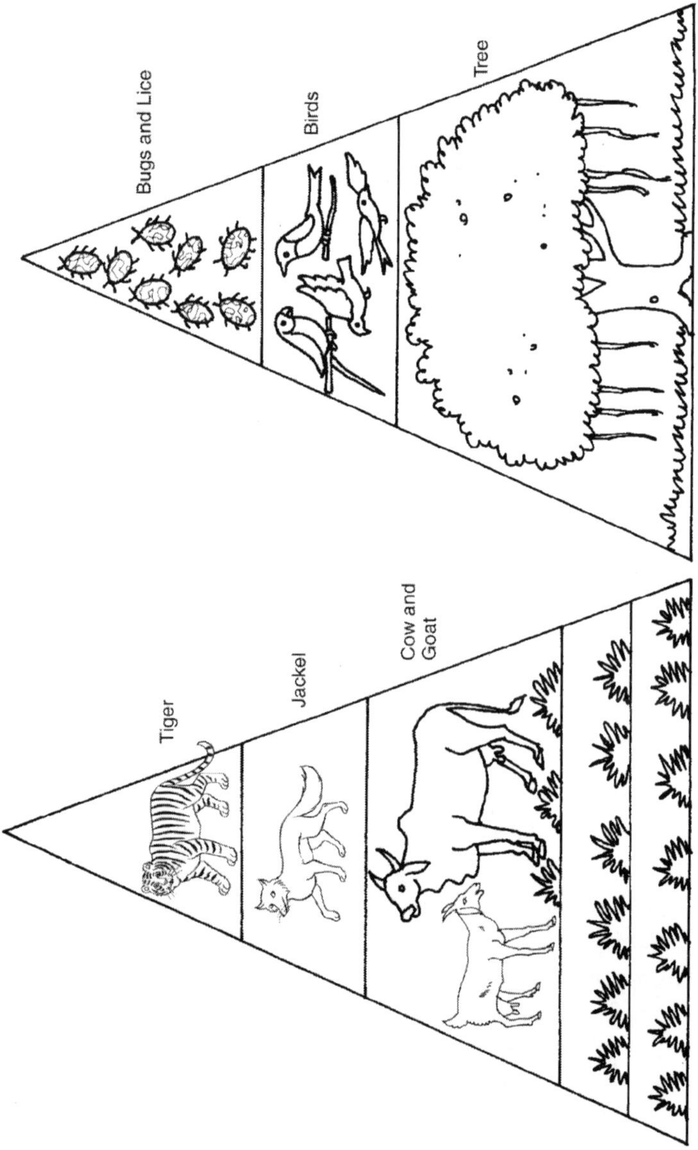

Fig.3.7 Upright pyramid of biomass (a) Grassland ecosystem (b) Tree ecosystem

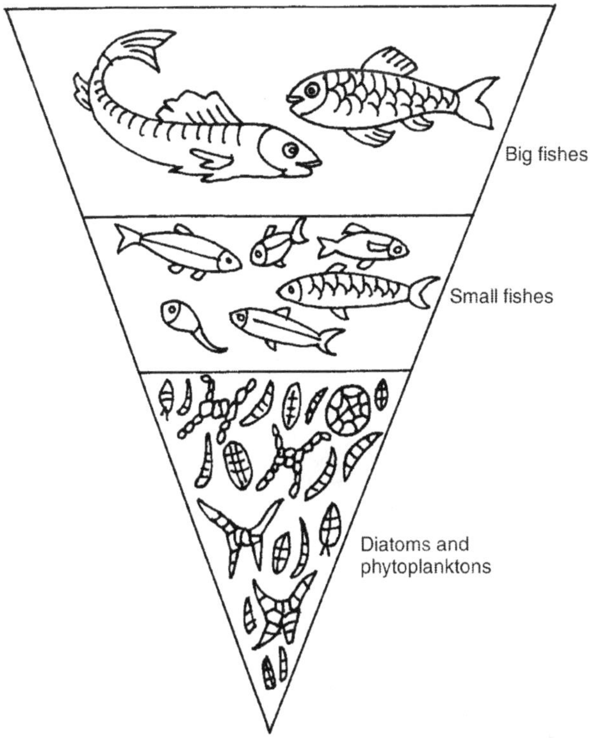

Fig.3.8 An inverted pyramid of biomass of a Pond ecosystem

from one trophic level to another trophic level decreases in the food chain as we reach to the top trophic level. Because large amount of the energy is lost in the form of heat, energy conversion is not 100 per cent. This principle confirms the second law of thermodynamics. Thus the principle of food chain and the two laws of thermodynamics can be best illustrated by energy flow models.

Energy flow in the ecosystem The behaviour of energy in ecosystem can be called as energy flow due to unidirectional flow of energy. From energetic point of view it is essential to understand in any ecosystem:

- the efficiency of the producers in absorption and fixation of solar energy.
- the use of fixed energy by consumers.
- the total input of energy in the form of food and its efficiency of assimilation.

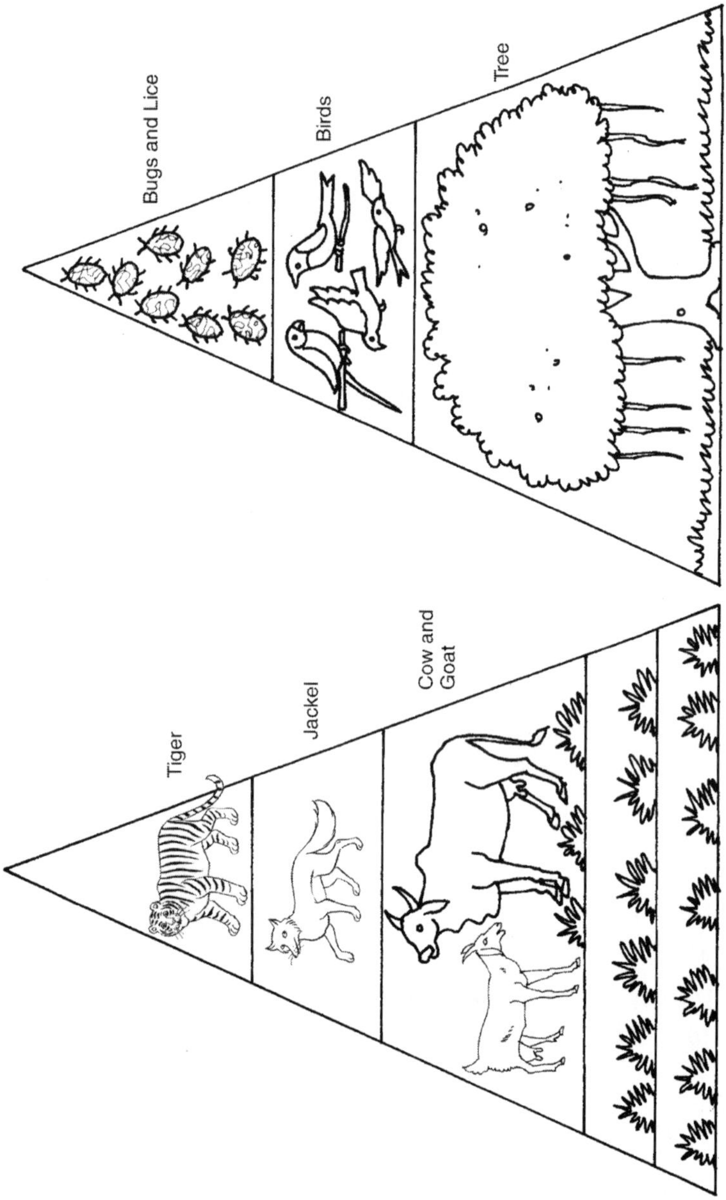

Fig.3.9 Upright pyramid of energy (a) Grassland ecosystem
(b) Aquatic ecosystem

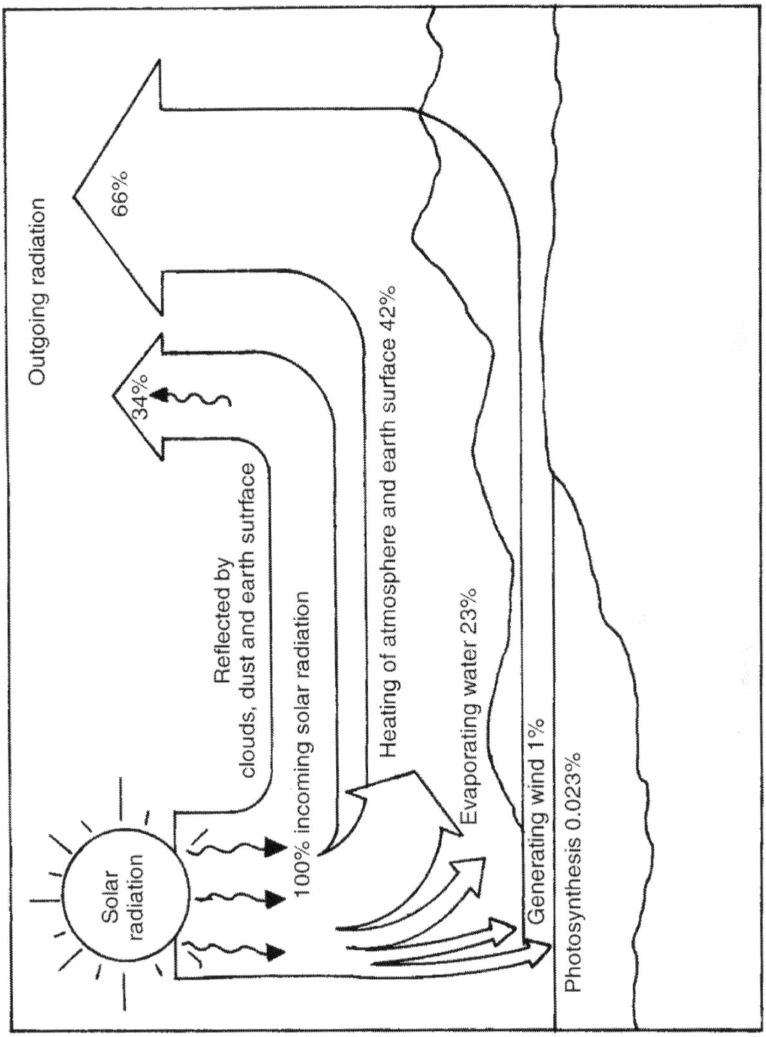

Fig.3.10 Flow of energy

- gross and net production.
- loss through respiration.
- loss through excretion.

Lindeman in 1942 gave the first energy flow diagram through different trophic levels.

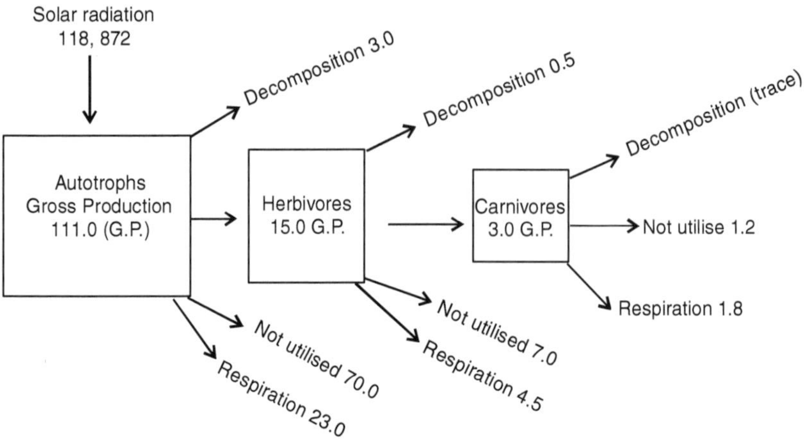

Fig.3.11 Energy flow diagram (modified from Lindeman 1942)

From the energy flow diagram three things become clear as shown in Figure 3.10.

- There is an unidirectional flow of energy.
- There is a progressive decrease in energy at each trophic level.
- There is a progressive improvement of transfer efficiency as energy passes along a food chain.

At each trophic level energy flow may follow three main paths:

- To the next trophic level.
- Through death to decomposers.
- Lost as a heat by way of respiration.

Behaviour of energy flow

The behaviour of energy in ecosystem can be called as energy flow due to unidirectional flow of energy. From energetic point of view it is essential to understand in any ecosystem

1. the efficiency of the producers in absorption and fixation of solar energy.
2. the use of fixed energy by consumers.

3. the total input of energy in the form of food and its efficiency of assimilation.
4. the gross net production.
5. loss through respiration.
6. loss through excretion.
7. loss through heat.

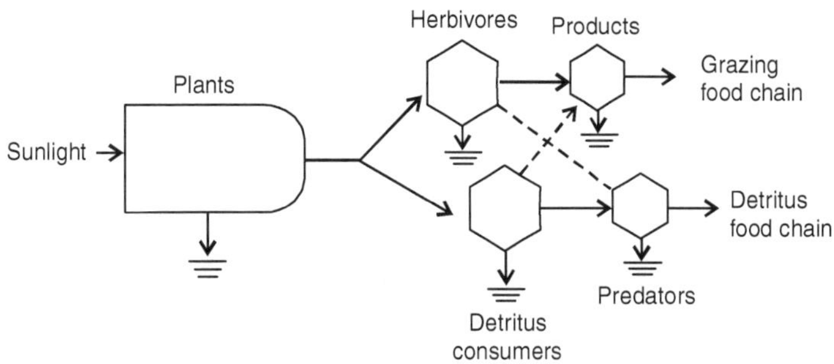

Fig.3.12 Y-shaped energy flow model (After E.P. Odum, 1983)

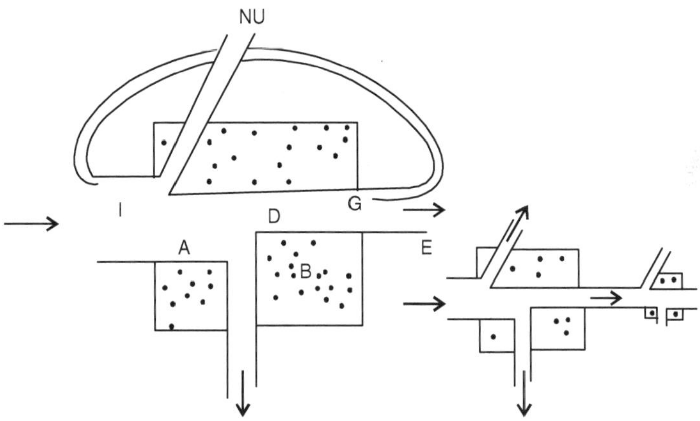

Fig.3.13 Component of Universal model of enrgy flow (After E.P. Odum 1968)

I – Input
NU – Not used
A – Assimilated energy
P – Production
R – Respiration
B – Biomass
G – Growth
S – Stored energy
E – Excreted energy

To understand energy flow in any ecosystem there are two types of models:

1. Single-channel or simplified energy flow model.
2. Two-channel or Y shaped energy flow model.

Single channel energy model

In this single channel energy model three things are important.

1. There is unidirectional flow of energy. The energy absorbed by herbivores passes to carnivores and so on. There is no reversion of energy. Energy moves progressively from one trophic level to another trophic level.
2. There is progressive decrease in energy from one trophic level to another trophic level. Large amount of energy utilized in metabolic activity, respiration and other remain unutilized or wasted.
3. There is progressive improvement of transfer efficiency as energy passes along a food chain.

At each trophic level energy flow may follow three main paths:

1. To the next trophic level
2. Through decomposers
3. Lost as heat by way of respiration

Solar energy falling on earth is 118,872 gcal/cm^2/yr. Out of which 118,761 gcal/cm^2/yr remains unused. Efficiency of energy capture by autotrophs is only 0.10 per cent. So the gross production by 1 autotroph is 111 gcal/cm^2/yr. Only 23 gcal/cm^2/yr is consumed in respiration by autotrophs and 70 gcal/cm^2/yr remains unutilized. Further it may be noted that 15 gcal/cm^2/yr is consumed by herbivores and 4.5 gcal/cm^2/yr is used for respiration. 0.5 g cal/cm^2/yr is utilized for decomposition at trophic level. Thus, more energy is lost in herbivores than in autotrophs. Again a large amount of energy remains utilized in the carnivores. About 3.0 gcal/cm^2/yr passes to carnivores. 6 per cent of the energy intake is consumed in metabolic reactions and only an insignificant amount is utilized for decomposition.

Simplified energy flow model

This model was proposed by Odum 1963. This model shows three trophic levels. Each trophic level and energy flow is shown by boxes and pipes. The boxes represent the biomass and the pipes show the flow of energy between living units. Total solar radiation from sun falling upon the green plants is 3,000 kcal. Out of which 1500 kcal is absorbed and only 15 kcal is converted at first trophic level. So primary production is only one per cent whereas secondary productivity is 10 per cent in the successive trophic levels. Hence, it is clear from the two models that there is a progressive decrease in energy flow at each trophic level. Thus longer the food chain, lesser would be the available food energy and shorter the food chain, more would be the availability of food energy. So this model clearly shows that energy flow is greatly reduced at each successive trophic level and at each trophic level with the transfer of energy, a large amount of energy is lost as heat or other form.

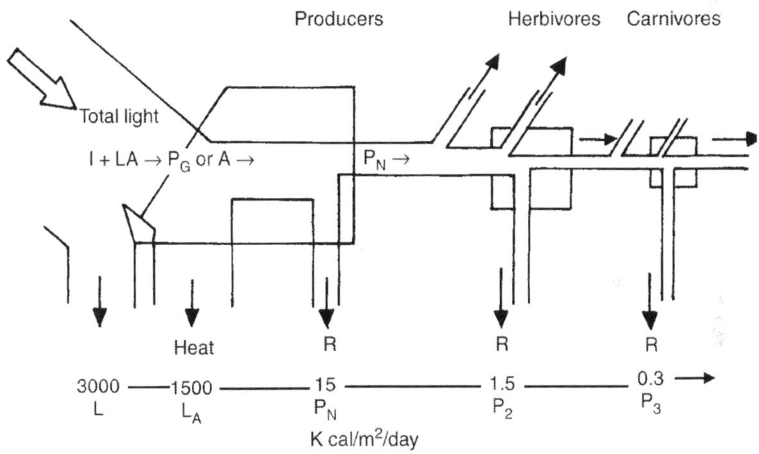

Fig.3.14 Simplified energy flow model (After E.P. Odum 1963)

I – I – Total energy input R – Respiration
L_A – Light absorbed by plant cover
P_G – Gross primary production
A – Total assimilation
P_N – Net primary production
P – Secondary production
NU – Energy not used
NA – Energy not assimilated by consumers.

Y-shaped or Two-channel energy flow model

This model was given by H.T. Odum in 1956. This model shows a common boundary, light, heat flows, import, export and storage of organic matter. Decomposers are placed in different boxes to separate the grazing and detritus food chains. In this model two food chains are sharply isolated. In marine ecosystem the grazing food chain is major pathway of energy flow whereas in the forest ecosystem detritus food chain is more important because in this 90 per cent of the net primary production is utilised. The distinction between the two food chains is of time lag between the direct consumption of living plants and ultimate utilization of dead organic matter. The Y shaped energy flow model is more realistic than the single channel energy flow model because

1. It confirms the basic stratified structure of ecosystem.
2. It indicates the isolation of grazing and detritus food chain both in time and space.
3. The micro-consumers differ greatly in size-metabolism relations.

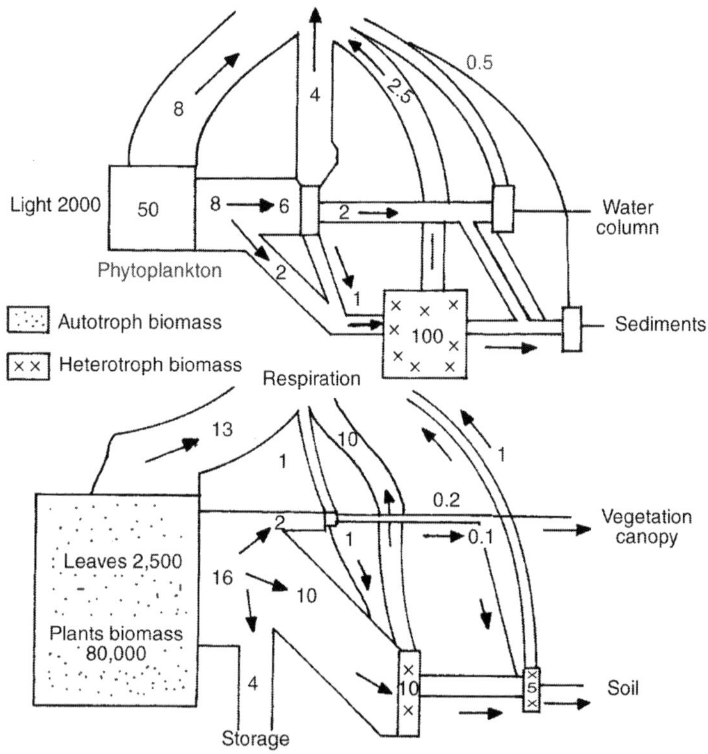

Fig.3.15 Y-shaped energy flow model showing separation of detrifus food chain (sediments and in soil) and grazing food chain (water column or vegetation canopy) (Modified from E.P. Odum 1963)

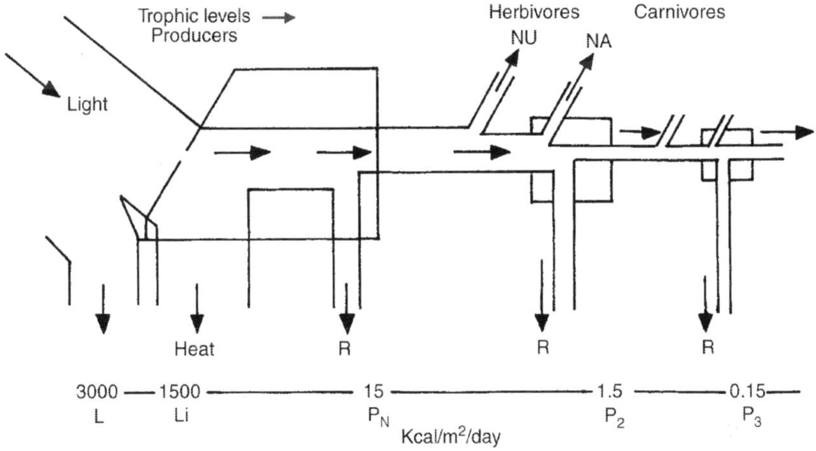

Fig.3.16 Energy flow diagram (E.P. Odum 1963)

L – Total light
P_N – Net production
R – Respiration
NA – Non-assimilated energy
P_G – Gross production
I – Energy intake
NU – Not utilised energy

Ecological Succession

Environment and communities are never stable. They keep on changing more or less regularly over time and space. They are never found stable with their physical environment. Environment always keeps on changing with time due to following factors:

1. Variations in climatic and physiographic factors
2. Activities of the communities' species

These two factors bring about drastic changes in the existing communities and lead to replacement by another community at the same place. This type of process continues till a stable community develops for a period of time on the same area. This process is known as **ecological succession**.

Succession means "coming in sequence". Hence, ecosystem progresses and leads to the evolution of a stable community called **ecological succession**. It can be defined as the occurrence of relatively definite sequence of communities over a period of time in the same area resulting in establishment of final complex community. E.P. Odum describes ecological succession with the help of following principles.

1. It is an orderly process of community development in which one species structure is replaced by another species with time.
2. It is directional and predictable process.
3. It results from the modification of the physical environment by the existing community. The physical environment determines the pattern, rate of change and the extent of development of community.
4. It results in a stable community in which maximum biomass and mutual functions between the organisms are maintained per unit energy flow.

The first community that inhabits an area is known as **pioneer community**. This community takes time to modify the environment for the entry of another community in that particular area. So the intermediate community is known as **seral community** of that area. The final community is the stable community and is known as **climax community**.

Trends of succession An ecological succession follows the given pattern:

1. There is a continuous change in the kind of plants and animals.
2. There is an increase in the diversity of species.
3. There is an increase in the organic matter and biomass in autotrophic succession whereas in heterotrophic succession the pattern is just reverse.
4. There is a decrease in net community production.

Causes of ecological succession Succession is a complete and natural process. There cannot be a single cause for the development. Mainly there are three causes which are as follows:

1. **Initiating or Initial Cause:** Initiating cause creates a bare area or destroy the existing organism in an area. There are two causes, namely, climatic and biotic. Climatic causes include erosion, deposits, wind and fire and so on. Biotic causes include the activity of the organisms.
2. **Continuing or Ecesis Cause:** This type of cause includes migration, aggregation, competition, and reaction. This causes successive waves of population as a result of changes in environment that mainly involve edaphic features of the area.

3. **Stabilising Cause:** This causes the stabilization of the community.

Types of succession These are some of the basic types of succession:

1. **Primary succession:** It starts from the parren substratum, hence are the first group of organisms which get established. These are known as primary colonisers or **primary community or pioneers**.
2. **Secondary succession:** It starts from the previously built substratum. But due to some external or internal disturbances it gets lost completely or partially. This is called secondary succession. This type of succession is more rapid.
3. **Autogenic succession:** *Auto* means self and *genic* means to generate. After the succession begins, community itself modifies its own environment and thus causes its replacement by new communities.
4. **Allogenic succession:** It occurs when there is replacement of existing community by the external agencies like drought, temperature, rainfall and erosion.

Sometimes succession is classified on the basis of changes in nutritional and energy contents. These are as follows:

1. **Autotrophic succession:** *Auto* means self and *trophic* means energy. This type of succession starts with autotrophic organisms. It begins with relatively rich in inorganic elements and energy. There is a gradual increase in organic matter and energy flow.
2. **Heterotrophic succession:** *Hetero* means different, *trophic* means energy. This type of succession starts with rich organic matter and energy content area. Here, progressive decrease in organic matter and energy content takes place as we proceed towards climax state.

In theory there are many other kinds of successions depending upon the environmental conditions. When succession starts in minimum amount of moisture then it is known as *xerosere*. When succession starts in adequate moisture conditions than it is known as *mesosere*. Similarly when the process begins in plenty of water then it is known as *hydrosere*.

Process of succession Autotrophic succession is completed through a sequential process. The process is as follows:

1. **Nudation:** This is the first step and involves creation of a bare area by a member of natural agencies such as erosion, land slides, floods, volcanic eruptions and epidemics and so on.
2. **Invasion:** It is the entry on the bare area through the biotic and abiotic mode. It involves three steps.
   (i) **Migration:** It involves the transfer of seeds and spores by air or water into a bare area.
   (ii) **Ecesis:** Once the species enter the bare area they must grow, establish and reproduce for successful establish-ment. This process depends upon the environmental factors of the area. Species adapted to the environmental conditions get established in the area.
   (iii) **Aggregation:** As a result of establishment, individuals of the species increase in number by reproduction and come close to each other. This process is known as *aggregation.*
3. **Competition:** The aggregation of species in a limited area results in competition for space and nutrition. This can be interspecific and intraspecific competition.
   (i) *Interspecific competition:* Competition between two different species.
   (ii) *Interaspecific competition:* Competition within the same species.
4. **Coaction:** Individuals of the species affect each other in many ways. This is known as *Coaction.* The species which are unable to compete are discarded or die in that area. Surviving species have high reproductive ability and have wide ecological amplitudes.
5. **Reaction:** The mechanism of the modification of the environment through the influence of living organism is known as *reaction.* Due to reaction, changes takes place in soil, water, light conditions and temperature of the area. Thus, the environmental conditions get modified in that area. The existing environment become unsuitable, for the community and gets replaced by another community. This another community is known as *Seral Community.*
6. **Stabilization:** It is the stage when the final community becomes more or less stabilized for a longer period of time. This final

community is not replaced and is known as climax community. This stage is known as *climax stage*.

Types of succession based on environmental conditions

Hydrosere: Hydrosere starts from water bodies such as ponds, lakes etc. In this, successive changes take place in animals and plants. Fresh water bodies have continuous intake of soil and organic matter through water. It starts with the colonisation of phytoplanktons and zooplanktons. These pioneer species are replaced by dominating plant and animal species with the accumulation of organic matter at the bottom of water body. At every stage, the original existing community affects the climatic and edaphic conditions of that area. The final stable community which terminates into a forest is known as climax stage.

Various stages of hydrosere are as given below

1. **Phytoplankton Stage:** These constitute the starting community. Bacteria, diatoms and blue green algae were the first coloniser in primitive time.
2. **Rooted Submerged Stage:** Decay of the body of Phytoplankten mixes with silt and creates soft mud in the pond which act as the habitat for plants like Hydrilla, Utricularia.
3. **Rooted Floating Stage:** By the death and decomposition of rooted submerged plants water level of pond becomes shallower and depth is near up to 6 feet. Dominating species are Nymphea, Trapa, Lemna, Salvinia and so on.
4. **Reed Swamp Stage:** Decay of the rooted floating stage plant builds the substratum for this stage. Roots remain in the water and shoots remain in the air. Dominating species Typha, Sagiitarius and so on.
5. **Seed Meadow Stage:** In this stage water level reduces and changed substratum mainly supports members of graminal and Cyperaceae family.
6. **Woodland Stage:** Now soci becomes drier and invaded by the shrub. Accumulation of humus and mineralisation of soil favours the free species.

7. **Forest Stage:** Forest stage is the climax community and supported by Almus, Acer and Quercus trees.

Successional seral stages and associated plant and animal species in hydrosere are depicted in Table 3.3.

Table 3.3 Succession in pond

Various Seral Stages	Plant Species	Animal Species
Initial Stage Pioneer Community	Phytoplanktons, cynobacteria, diatoms etc.	Zooplanktons, Eco-gland paramecium etc.
Rooted submerged stage seral community	Potamogeton, Hydrilla, Myriophyllum, Vallisneria	Flies, Daphania, Cyclops, Ascellas, Gannarus
Rooted Floating Stage Seral Community	Nelumbo, Trapa, Nymphaea, Aponogeton, Free floating plants	Beetles, Hydra, Arthropods, Fishes, Turtles, Water snakes, Frogs
Reed-swamp stage Seral community	Specis of Scirpus Typha, Phragmites etc.	Flies, Scorpions, Beeltes and Lung breathing animals
Sedge meadow Stage Seral Community	Carex, Cyperus, Juncus, Eleocharis etc.	Beetles etc.
Wood land stage	Salix, Cornus, Populus and Almus etc.	Birds, insects, rabbits etc.
Forest Stage Climax Community	Herbs, Shrubs, Trees, parasites, Saprophyte, Detrivores	Herbivores and carnivores

Lithosere—A Xerosere on rock: Xerosere starts on bare rock surface. The rock surface is characterised by lack of organic matter, deficiency of water and high surface temperature. The pioneers to colonise are crustose type of lichen and through a series of successive seral stages, the succession terminates into a forest which is a climax community. As the crustose lichens grow, environment becomes less favorable and replaced by foliose lichen. This foliose lichen becomes less favorable for the growth due to addition of organic matter and replaced by mosses. Similarly, mosses are replaced by herbs, shrubs and finally the climax forest in course of time.

Various states and main species representing the stage are discussed below.

1. **Crustose Lichen Stage:** Substratum is deficient in water and organic matter. Rhizocarpass, Rinodina are mainly present and secrete acid by which weathering of rocks takes place.
2. **Foliose Lichens Stage:** Death and decay of crustose lichen mix with the minerals of rocks and favours foliose lichen. Dermato carpon and Parmelia includes foliose lichen.
3. **Moss Stage:** Thin layer of soil acts as the substratum for the polytrichum, tortula.
4. **Herb Stage:** Decay of moss plants add more minerals to soil and this change favours shallow rooted grass like Poa, Festuca and so on.
5. **Shrub Stage:** In this stage much accumulation of soil supports phytocarpus like shrub plants.
6. **Forest Stage:** Forest stage has enough soil to support Mesophytic trees.

Different seral stages and associated plants and animal species in a lithosere are depicted in Table 3.4.

Table 3.4 Lithosere—Succession on Rock

Different seral stage	Plant species	Animal species
Crustose lichens	Rinodina, Lecanora Rhizocarpon etc.	Spiders, ant, unites etc.
Foliose lichens	Dermatocarpon and Permelia etc.	Spiders and ants etc.
Moss stage	Torula, Polytrichum and Grimmia etc.	Ant, spiders etc.
Herb stage	Eleusine, Aristida, Solidago, Festuca and Poa etc.	Nematodes, insects, ant, spiders, larvae etc.
Shrub stage	Phyto corpus and Rhus etc.	Arthropoda, Nematodes Annelids etc.
Forest stage	Herbs, shrubs and trees	Carnivores, herbivores, saprophytes, parasites and detritivores etc.

Heterotrophic succession: This type of succession is heterotrophic in nature and starts from a habitat rich in organic matter content. This succession involves the opposite progression, from an initial condition of high energy content to a final stage of low energy content.

ECOSYSTEM IN VARIOUS HABITATS

Forest Ecosystem

Forests are the natural plant communities and cover about 40 per cent of the land. Indian forest covers about 22.74 per cent of the total land area of country. Coniferous, deciduous and tropical rain forests are the main dominating forest cover of the earth. Coniferous forest have short summers, long and cold winters. Tropical forests experience high humidity and heavy rainfall. In India forests are about 22.74 per cent of the total land area.

Types of forest ecosystem

1. **Tropical rain forest:** It is found in Amazon basin of South America and experiences high humidity and rainfall.
2. **Savanna:** Here forest merge with grasslands.
3. **Boreal forests:** It covers the northern part of globe. Pines and Cedars are the dominating trees.
4. **Temperate rainforests:** It has high moisture, humidity and warm temperature and dominated by cedar, spruce and fir trees with lavish vegetation.
5. **Temperate evergreen forest:** It has long winters with heavy snow and is dominated by pines and fir trees.
6. **Temperate woodland:** It has small trees with well-developed grasslands, shrubs, dominated by pines and junipers.
7. **Temperate deciduous forest:** The summers experiences moderate rainfall, moderate temperature, cold winters and dominated by deciduous trees.

Structure of ecosystem It consists of two components abiotic and biotic.

1. **Abiotic component:** This component consist of inorganic substances, organic substances, climatic and edaphic factors. Light and moisture conditions are variable due to complex stratification. Minerals and dead organic debris are also present in the forest.
2. **Biotic component:** This component consists of living organisms such as producers, consumers and decomposers.
   (i) *Producers:* These are autotrophs and include trees, shrubs and ground vegetation. Trees show a variety of species in a

forest. Tropical forests are characterised by evergreen trees with leathary leaves. In India, tropical forests are dominated by red sanders, bijasal, anjan, teak, axlewood and paurel. Thorn forest axlewood, sandal wood, neem, acacia senegal, neem, palas, ak, ranju, khair etc. Subtropical forest have *machilus sp.*, jamun, *mehiosma sp.*, Chirpine and so on. Tropical moist deciduous forests have semul, sandal, sal, teak, dhaman species, haldu etc. Temperate forest have conifers, oaks and grasses.

(ii) *Consumers:* These feed on producer and include primary consumers, secondary consumers and tertiary consumers.
Primary Consumers: These include herbivores that feed on leaves, shrubs and vegetation. Examples are beetles, ant, leaf hoppers, spiders and bugs etc. Mice, squirrels, rabbits, nilgai, deer, feed on fruit of producers.
Secondary consumers: These are carnivores such as foxes, birds, lizards and snakes.
Tertiary consumers: These are top carnivores that feed on secondary consumers and herbivores. Examples are Tiger and Lion.

(iii) *Decomposers:* These include micro-organisms such as Aspergillus, Fusaruim, Trichoderma, Comrinus, *Alternaria sp.*, Bacteria and Actinomycetes etc. In forest ecosystem, decomposition is faster in tropical and subtropical areas than in the temperate areas.

Functions of forest ecosystem Forest ecosystem helps in numeral cycling and addition of elements in nitrogen fixation etc.

Grassland Ecosystem

Grassland occupies about 19 per cent of Earth's surface. It has high rate of evaporation and experiences severe droughts periodically. Rainfall is between 10 to 30 inches per year. Temperate grasslands of the world are known by different names such as Steppes of Europe and Asia, Veldts of Africa, Tussocks of New Zealand, Priaries of Canada and North America and Pampas of South America.

Types of grasslands

Tropical Grasslands: It has maximum photosynthetic rate. Plant material has low crude protein and high crude fibre. These are also known as tropical savannah. These are located in Africa, Australia and South America with an area 15×10^6 sq. km. These type of grasslands have wet seasons alternating with long dry seasons. The rainfall varies from 100 to 150 cm. Flora includes grasses, shrubs and a few trees whereas, fauna includes zebra, giraff, elephant, antelopes, tigers, kangaroos and lion.

Jame grasslands: These are cleared forest regions and are utilised for food as livestock.

Tall-grass Praries grassland: These grasslands are a narrow belt running from north to south along the border of Canada and North America. It has continuously changing series of species.

Mixed Prairie grassland: It experiences high precipitation. It is dominated by mid grasses in moist years and short grasses in dry years.

Short grass plains grassland: It is characterised by light rainfall, humidity, high wind and evaporation. It is dominated by short grasses and followed by mid grasses.

Desert grassland: It has hot, dry climate and dominated by grasses and Savannas. The grasslands are dominated by fauna such as antelopes, zebra, bighorn sheep, deer, rats, mices, dogs, squirrels, wolves, birds, lion, tigers, giraff and elephant.

Grasslands are divided into three strata–Roots, ground layer and the herbaceous layer.

Structure of grassland ecosystem It consist of abiotic and biotic components.

1. **Abiotic component:** It includes the inorganic and organic compounds. It also contains nutrients in the soil and atmosphere such as carbon, oxygen, hydrogen, nitrogen, phosphorus, sulphur which comes from water, nitrates, carbon dioxide, phosphates etc. Climatic and edaphic factors also constitute the abiotic component.
2. **Biotic component:** It consists of components—producers consumers and decomposers.

   Producer: These include grasses, herbs and shrubs. Main grass species are sporobolus, setaria, panicum, andropogon, cynodon, dicanthium etc.

Consumers: These includes primary and secondary and tertiary consumers.

Primary consumers: These are the herbivores and include fauna like rabbits, mice, rat, hare, deer, goat, cow etc. Termites and insects are also present that feed on leaves of grasses.

Secondary consumers: These are the carnivores and include lizard, fox, snake, birds and frogs.

Tertiary consumers: These are the carnivore and mainly involve predatory birds and hawks.

Decomposers: These are fungia, bacteria and actinomycetes that feed on dead matter. These recycle inorganic substances back to soil and atmosphere.

Desert Ecosystem

Deserts occupy about 1/7th of land surface of the Earth. Deserts are those regions where evaporation exceeds rainfall. The annual rainfall is less than 25 centimeters. Hence, deserts experience low rainfall and high evaporation with different temperatures. Deserts have harsh environment. Plant and animals evolve ways to circumvent aridity and temperature. The number of species vary in the ecosystem due to extreme temperature and water. These ecosystems result from climatic pattern of the Earth. These have low net production, opportunistic grazing herbivores and carnivores.

The main deserts are Sahara and Kalahari desert in North and South Africa, Great Western desert in the South–western USA, Tibet, Gobi and Thar deserts in Asia, Central-western Australia desert, Atacama desert of South America. In desert, days are hot and nights are cold. Soil is sandy and hot.

Deserts are of two types—hot deserts such as Sahara and Thar, cold deserts such as Gobi desert. Flora and fauna is poor. Desert plants are hard such as Acacia, Cacti, Euphorbias and various types of succulants whereas fauna includes locusts, spiders, lizards, birds, squirrel, foxes, mongoose, rats, kangaroo, desert cats and so on.

Structure of ecosystem Structure of ecosystem: abiotic and biotic components.

1. **Abiotic components:** These include inorganic and organic substances in soil and atmosphere.

2. **Biotic component:** The various biotic components are producers, consumers and decomposer.

   Producers: These are the autotrophs and include hard grasses, shrubs, bushes, some species of trees and some succuluents. Leaves, branches, stems and root system are modified to reduce transpiration. Few succulants such as cacti are present. Lower plants like lichen and xerophytic mosses are also present in desert ecosystem.

   Consumers: They feed on autotrophs and include animals such as insects, spiders, grazing, burrowing, browsing animals, snakes, birds and lizards. These animals are able to live under xeric conditions.

   Decomposers: These are few due to poor vegetation and hence less organic matter. Fungi and bacteria are the organisms in desert ecosystem and are thermophilic in nature.

Aquatic Ecosystem

Aquatic ecosystem includes fresh water and marine water ecosystems.

Fresh water ecosystem includes lentic and lotic systems. Marine water ecosystem includes open ocean and coastal ecosystems.

Pond Ecosystem

Pond ecosystem is an example of lentic system (standing water). This ecosystem is rich in vegetation, micro-organisms, plants and animals. This system is a self-regulatory fresh water ecosystem. It has abundant inorganic and organic substances. Pond ecosystems are shallow water bodies. Pond water bodies have various zonations. These zones are as follows:

1. **Litorial zone:** This is a shallow water zone and includes variety of free floating producers and rooted plants. Here the penetration of light is very high. Hence, it is nutrient rich area. Fauna includes snakes, snails and frogs.
2. **Limnetic zone:** This is open water zone. Here penetration of light is sufficient for photosynthesis. But light penetration is less than the litorial zone. This region includes a variety of phytoplanktons, zooplanktons and fishes.

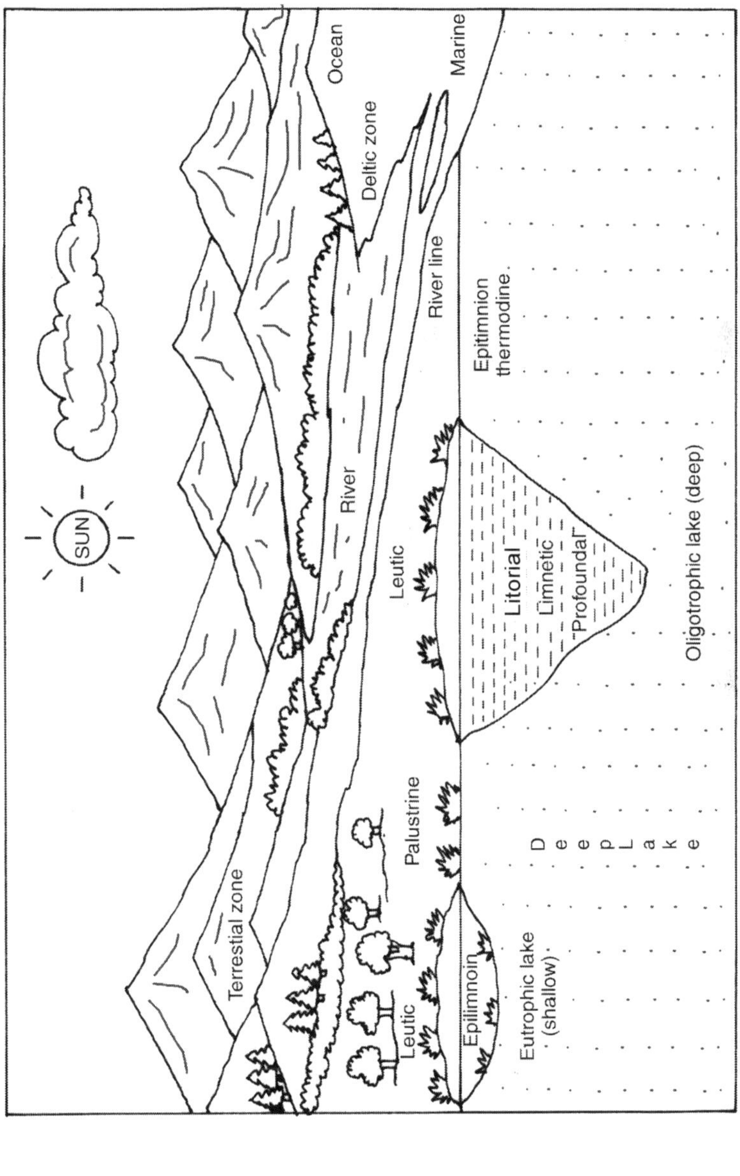

Fig.3.17 Zonation of lentic and lotic water bodies

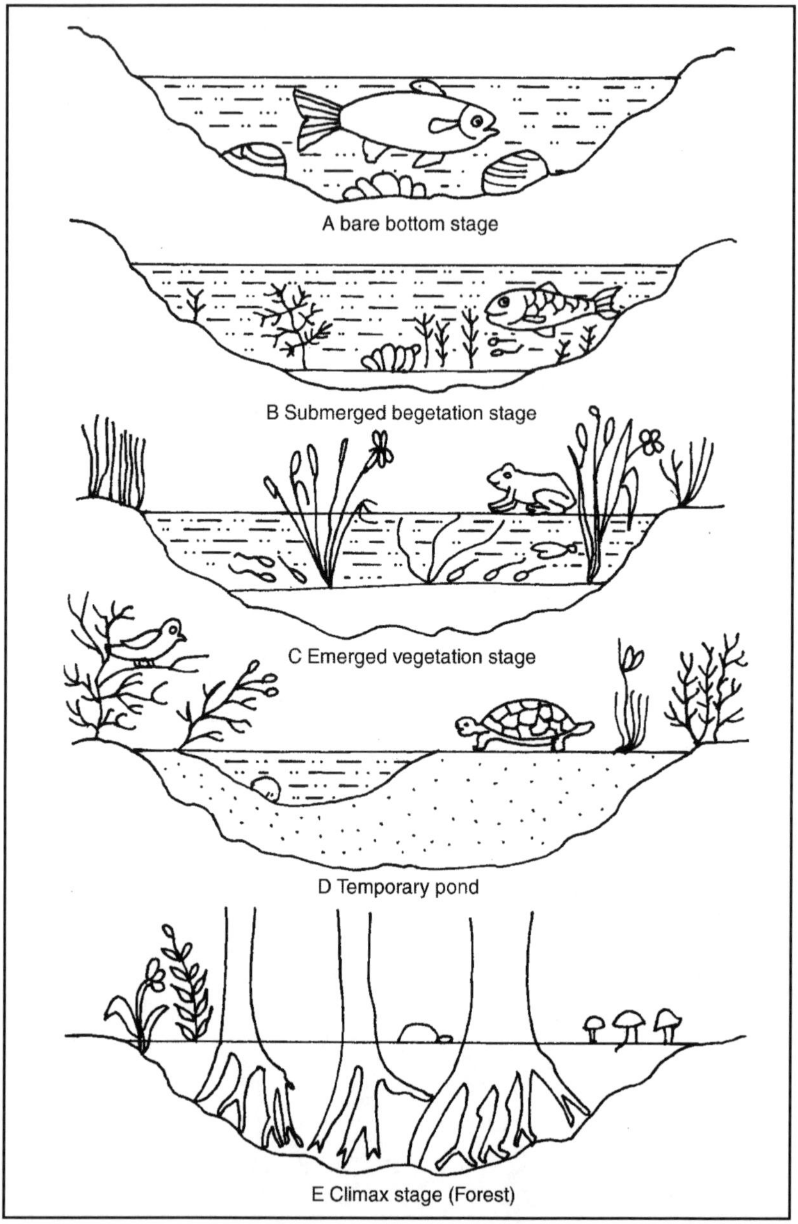

Fig.3.18 Ecological succession from a pond to a climax forest (after Cockrum and Mc Caulry, 1965)

3. **Profoundal zone:** This is a deep zone. Here biological activity is intense. Decomposition of organic matter takes place in this zone. Fishes are also present in this region and are adapted to cold water.

Structure of pond ecosystem It includes abiotic and biotic components.

1. **Abiotic component:** The abiotic component consists of inorganic and organic components. Only a small amount is found in soluble form and a large portion of these compounds are present in bottom sediments and in the organisms. Climatic factor is also a part of abiotic component. It helps in regulating the functioning of ecosystem.
2. **Biotic component:** Biotic component includes producers and consumers.

   Producers: It includes phytoplanktons, such as anabaena, closterium, volvox, oscillatoria, Eudorina etc. Filamentous algae, Submerged plants and floating plants.

   Consumers: It includes primary consumers and secondary consumers.

   Primary consumers: These are also known as herbivores. Insect larvae are the primary consumers.

   Secondary consumers: These are also known as carnivores. It involves zooplanktons such as Mysids, Lyclops.

   Dwelling animals such as annelids and molluscs contribute to secondary consumers. These in turn are eaten by large fishes called as tertiary consumers.

   Decomposers: It is an important component of ecosystem. Bacteria and fungi are present at the bottom of the pond water. These micro-organisms decompose the dead bodies and release the locked inorganic substances into the pond water for reuse by the producers.

Lake Ecosystem

Lakes are inland depressions. These are the standing water bodies formed by flowing water, precipitations and runoff water. The depth ranges from few feet to 5000 feet. The gradation of oxygen, light and temperature affect the life in the lake. Circulation of nutrients and distribution of organisms are influenced by currents. Lakes have different zonations like ponds and provide a number of ecological niches and habitats to the organisms. Lakes are of two types, oligotrophic and entrotrophic lakes. Oligotrophic lakes are nutrient deficient whereas cartrophic lakes are nutrient rich.

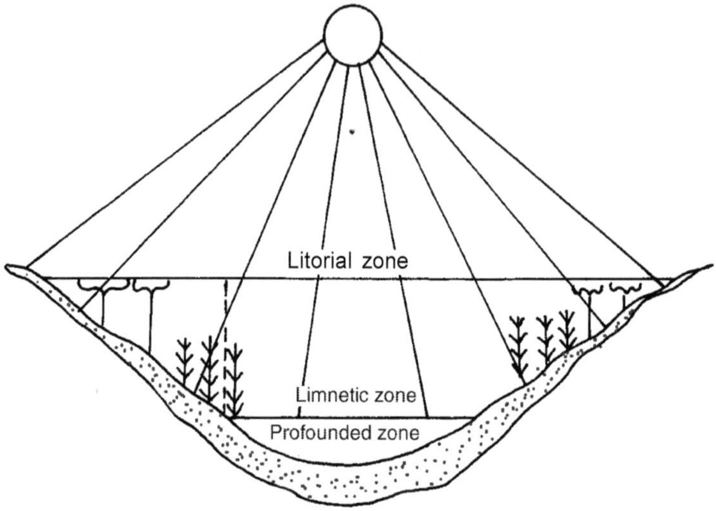

Fig.3.19 Different zones of fresh water pond (After Odum, 1971)

Structure of lakes Lakes consist of abiotic and biotic component.

1. **Abiotic component:** It includes light, heat, inorganic and organic compounds such as carbon dioxide, oxygen, nitrogen, phosphorus, calcium and amino acids etc. Climatic factors are also part of abiotic components.
2. **Biotic components :** It includes producers, consumers and decomposers.

   Producers: Producers are the autotrophs and include phytoplanktons such as diatoms (Navicula, Cyclotella) blue green algae, (Microcystis, Oscillatona) green algae, (Cosmarium, Staurastreum) and halophytic flagellatis and dinophycean flagellates.

   Consumers: They are heterotrophs and depend on producers for the food. Most of the consumers depend on producers for the food.

Most of the consumers are herbivores. Here consumers include oligocheates, Autodrilus Pluriseta, Branchiura Sowerbyi, Hirudinea species, molluscs and crustaceaus etc.

Decomposers: These are also known as micro-consumers because they absorb only a small amount of decomposed organic matter. These includes bacteria and fungia that feed on dead organic matter of consumer. They also play an important role in cycling of nutrients.

Marine Ecosystem

Marine ecosystem involves mainly oceans. This ecosystem covers about 70 per cent of Earth surface. Ocean ecosystem is more stable than the fresh water ecosystem in terms of chemical composition due to salinity, dissolved oxygen content, light and temperature. Oceans are divided into three regions, namely, photic, aphotic and abyssal region. Photic region is the upper portion where light penetration is maximum and includes phytoplanktons, zooplanktons and nektons as shown in figure. Aphotic

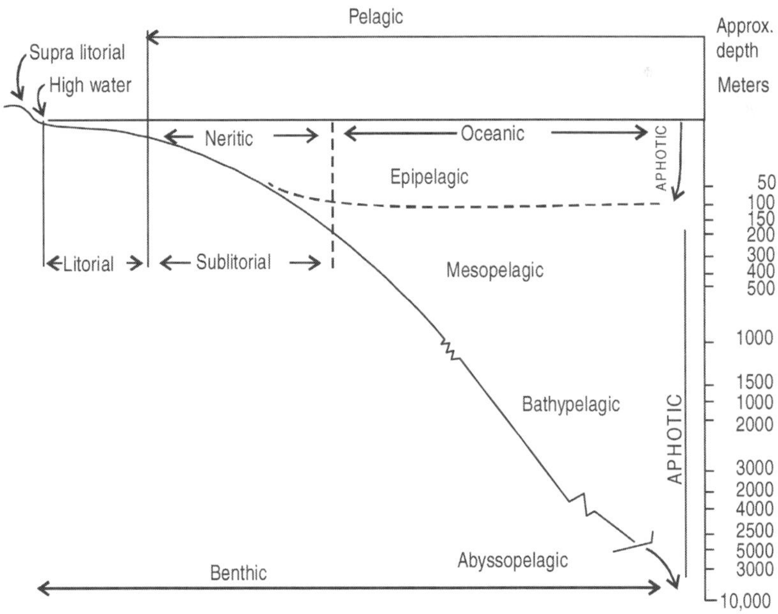

Fig.3.20 Classification of marine environment (after Hedgpeth, 1967)

region is the middle portion of the ocean and includes only nektons. Abyssal region is the lower portion and predators inhabit it.

Structure of oceans It consists of abiotic and biotic components.

1. **Abiotic components:** It includes inorganic and organic substances. Climatic factors are also part of abiotic components.
2. **Biotic component:** It includes producers, consumers and decomposers.

   Producers: It include autotrophs. Producers are phytoplankton such as dinoflagellates, diatoms and algae. In case of algae, phaeophyaceae and rhodophyceae are the dominating ones. All contributes to primary production.

   Consumers: These are the heterotrophs macro-consumers. They depend on primary producers for their nutrition.

   Primary consumers: These include herbivores. They feed on producers, mainly, molluscs, fish and crustaceans etc.

   Secondary consumers: These are the carnivores. They feed on herbivores. Examples of secondary consumers are fish such as shad, mackerel and herring.

   Tertiary consumers: These are also the carnivores. They feed on carnivores of secondary consumers. These mainly involve fishes such as Hadlock, cod etc.

   Decomposers: These are bacteria and fungi. They decompose dead organic matter of producer. They also help in cycling of nutrients by decomposition.

Estuaries

Estuaries are semi-enclosed coastal water bodies connected to oceans. It is a transition zone between fresh water and marine water. Estuaries have communities of endemic species. These are breeding grounds of the species. When mouth of a river opens into sea-water it is known as estuary. These water bodies are fertile in nature. This region is rich in carbon dioxide, oxygen and light. These areas are rich in species diversity.

Structure of estuaries

1. **Abiotic component:** These include inorganic, organic substance and climatic factors.

2. **Biotic component:** It includes following components:
   Producers: These are autotrophs and mainly include planktons, nektons and rooted plants.
   Consumers: There are hetrotrophs and include zooplanktons, small and large fishes.
   Decomposers: These feed on dead organic matter and include clams, bacteria, fungi, prawns, crabs, snakes, rays, snails.

Rivers

Rivers include moving water. They are variable and ranging in size from Ganga, Yamuna, Kalinadi, Sutluj, Gomti and so on. A river varies considerably over its length, as it changes from a mountain brook to a large river.

1. **Abiotic component:** These include inorganic, organic substance and climatic factors.
2. **Biotic component:** Biotic components of river are as follows:
   Producers: Producers includes phytoplanktons, diatoms, blue green algae, green algae and water moss.
   Consumers: Consumers includes *Cephalopteryx, Helodes, Phalacrocera, Gammarus* etc.
   Decomposers: It includes, bacteria and fungi which feed on dead organic matter.

Stream

Streams are open ecosystem with high-energy input coming from outside sources. Streams may begin as outlets of ponds or lakes or they may arise from springs and seepage areas. Temperature affects the composition of stream community, shallow water streams are not good for many organisms. Energy flow in a stream is poor as compared to forest ecosystem.

Types of streams These are of two types:

1. **Fast streams:** It has high velocity of water and makes excellent trout nurseries.
2. **Slow streams:** It has low velocity of water, silt and decaying organic matter accumulates on the bottom.

Structure of streams involves following components

1. **Abiotic component:** These include inorganic, organic substance and climatic factors.
2. **Biotic component:** Biotic components of streams are as follows:
   Producers: It includes phytoplanktons like diatoms, blue green algae and green algae like cladophora, *ulothrix* and water moss.
   Consumers: It includes *Cephalopteryx*, *Helodes*, *Phalacrocera*, *Gammarus* etc. Fishes are also present.

REVIEW QUESTIONS

Short Type Questions

1. Who coined the term ecosystem?
2. Define the term ecosystem.
3. What is ecology?
4. What are the characteristics of an ecosystem?
5. Name different types of ecosystem.
6. Name different types of structural components of ecosystem.
7. What are the functional aspects of ecosystem?
8. Define food chain.
9. Name different types of food chains.
10. What is food web?
11. What are ecological pyramids?
12. Name different types of ecological pyramids.
13. Define ecological succession.
14. Name different types of ecological succession.
15. Name different types of ecosystems.

Descriptive Questions

16. Discuss the structure and function of an ecosystem.
17. What is energy flow? Discuss in detail the different types of energy flow model.
18. Explain in detail the process of ecological succession with suitable examples.
19. What are the causes of ecological succession?
20. Discuss in detail the fresh water and marine water ecosystem with suitable examples.
21. Discuss the structure and function of forest, grassland and desert ecosystem.

4
Biodiversity

CHAPTER OBJECTIVES

- Introduction
- Bio-geographical Classification of India
- Values of Biodiversity
- Biodiversity at Global, National and Local Level
- India as Megadiversity Nation
- Hot Spots of Biodiversity
- Threats to Biodiversity
- Conservation of Biodiversity

INTRODUCTION

The word biodiversity was used for the first time in Washington in a Symposium in 1986. Biodiversity refers to the number, variety and variability of all life forms inhabiting Earth. For example, prokaryotes, eukaryotes, wild plants, animals, micro-organisms, domesticated animals, cultivated plants and even genetic materials like seeds and germ plasma are a part of diverse fauna and flora. The variety and variability of organisms and ecosystems is called **biological diversity**.

In simplest form, biodiversity is defined as the sum total of species richness. Biodiversity is the shortened form of two words "Biological" and "Diversity". The meaning of these two words is that life on Earth has many levels of diversity.

The richness of biodiversity in India is largely due to geographical position. It possesses all kinds of ecosystems. There are about 30 million species present on planet Earth. Out of which only 1.5 millions have been identified till now and over millions remain unidentified. There are about 259,000 green plants, and 65,000 species of animals.

Biodiversity exists at three different levels:

1. Species diversity
2. Ecosystem diversity
3. Genetic diversity

Species Diversity: Species diversity is the variety of species within a region. It can be measured at the species level. For example, *Panthera leopersica* (lion), *Panthera tigris* (big cats) and *Pantheria uncia* (snow leopard). All these belong to same genera but are all different species. So, species are reflected by the morphological, physiological and genetic features. India is a home of about 2,00,000 species of living organisms. Of the total species, several are endemic in India.

Ecosystem diversity: Ecosystem diversity is related to the different types of ecosystems and habitats. It is the number of species in a community of organisms, such as coasts, deserts, lakes, and estuaries. Ecosystem diversity is difficult to measure since the boundaries of the communities are elusive.

Genetic diversity: Genetic diversity refers to the diversity of genetic poll within a species. Hence, it is concerned with the variation of genes within the species. Due to variations in genes, diversity in characters occurs within the same species and gives rise to varieties and races. For example, there are different "races" of human beings like Negrito, Nordic and Mongoloid and similarly there are different varieties of gram and rice.

Bio-geographical Classification of India

Bio-geography is the study of biota with regard to their origin, environmental interrelationships and distribution. Bio-geographical areas in India with respect to diversity are as follows:

Botanical survey of India (BSI) has divided Indian flora into following phyto-geographical regions.

North Western Himalayas: It is located in the sub-tropical dry ever green pine forests. It ranges from central region, in the foothills of the North Western Himalayas. It mainly includes dry and moist alpine scrubs, and coniferous trees.

Eastern Himalaya: This region is rich in biodiversity due to warm and wet climatic conditions. This area includes tropical semi- evergreen forests, subtropical hill forests, moist temperate forests, dry temperate forests and moist alpine scrubs.

Western Arid Regions: This region has four types of habitat conditions. (i) saline depressions (ii) rocky, (iii) sand dunes (iv) sandy plains.

Sand dunes and plains have very sparse vegetation and bushes. Saline depressions includes *Salvadora* and *Tamarix*. Rocky area includes *Zizypus, Cassia, Calotropis, Indigofera* etc.

Gangetic Plains: This area includes east Rajasthan, Uttar Pradesh, Bihar and Bengal. It includes tropical moist and dry deciduous forests and tall grasses.

Eastern India: This area includes the plain regions of Arunachal Pradesh and six eastern states. It includes tropical wet evergreen forests, tropical semi-evergreen forests, tropical moist deciduous forests, subtropical broad leaves hill forests and subtropical pine forests.

Deccan Plateau: This area covers eastern, western ghats and south of Vindhyanchal mountains. It has a triangular shape. It includes tropical rain forests, tropical dry deciduous forests, and moist deciduous forests.

Western Ghats: This area includes tropical moist deciduous forests, tropical semi-evergreen forests, tropical wet evergreen forests and subtropical broad leaved hill forests. This area is prone to endemism.

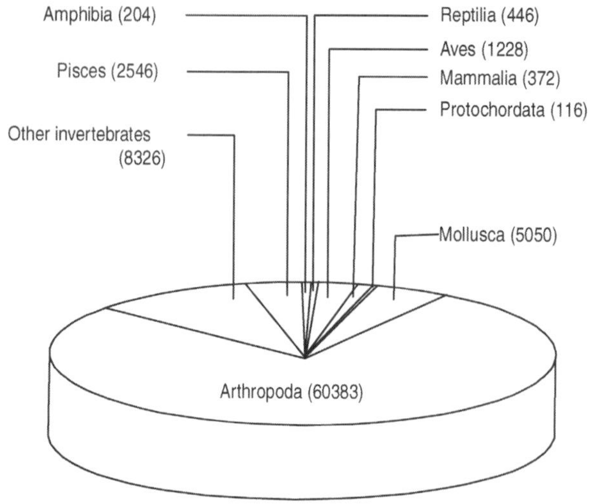

Fig.4.1 The distribution of recorded (81,251) Indian fauna

Source: Anon 1997, National Action plan on Biodiversity, Ministry of enviornment and forests, Government of India, New Delhi, p. 33

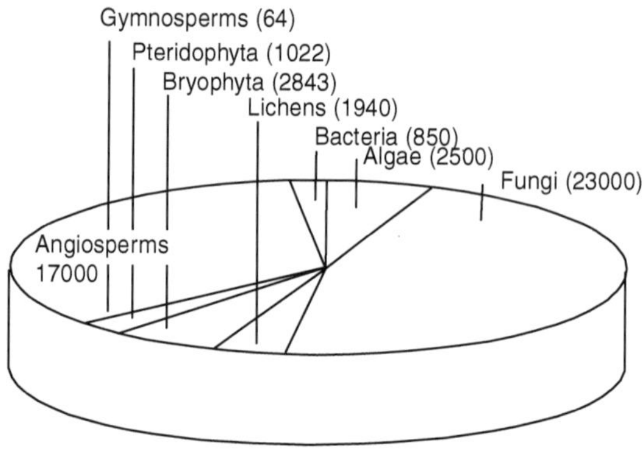

Fig.4.2 Distribution of recorded (49,219) Indian flora

Source: Anon 1997, National Action plan on Biodiversity, Ministry of enviornment and forests, Government of India, New Delhi, p. 32

Eastern Ghats: This area is poor in biodiversity as compared to western ghats. It is dominated by tropical dry deciduous forests.

Andaman and Nicobar Islands: These islands include tropical wet evergreen, semi-evergreen and moist deciduous forests.

Biogeographical Areas with Respect to Faunal Diversity

There are about ten biogeographical regions in India with respect to faunal diversity which are as follows:

1. **Trans-Himalaya:** It covers about 186,200 sq km in Ladakh and Lahulspiti. It has two protected areas. The faunal community includes goat, sheep, ass, leopard, wolf, marbled cat, pipa and marmot and black-necked crane.
2. **Himalaya:** It extends up to 236,300 sq km area. It has 56 protected areas. Faunal diversity includes markhor, western *tragopan*, snow leopard, brown beer, snowcock, husk deer, wild bear, sambar, muntjue.
3. **Desert (Kutch and Thar):** It extends upto 225,00 sq km area out of which 89 sq.km. area is protected. Faunal diversity includes sand grouse species, wild ass, desert cat and desert fox.
4. **Semi-arid:** It extends upto 508,000 sq km area. It has 52 protected areas. The lion of Gir is an endemic animal. The semi-arid region also includes the tiger reserves of Sariska and Ranthambhor.
5. **Western ghats:** It extends about 159,000 sq km area. It has 44 protected areas. It contains about 4000 plants species. Faunal diversity includes amphibian species dominance.
6. **Deccan peninsula:** It extends upto 1,421,000 sq km area. It has 115 protected areas. The faunal diversity includes sloth bear, leopard, chital, wild bear, barasingha, wild elephants, wild buffalo, and gair sambar.
7. **Gangetic plains:** It extends up to 359,400 sq.km. area. It has 45 protected areas. The faunal diversity consists of barasingha, blackbuck, bengal florican, turtle, crocodile, rhino and elephants.
8. **Marine coast:** This area consists of mangrove vegetation. The faunal diversity consists of inshore dolphin, estuarine and marine turtles, salt water and estuarine crocodile and tiger.
9. **North east:** It extends upto 1880 sq.km. area. It has 17 protected areas. The faunal diversity includes rhino, buffalo, elephant, swamp deer, hispid hare and hog deer.
10. **Indian islands:** It extends upto 708 sq.km. area. It has hundred protected areas. It consists of 2200 species of higher plants and 225 species of avifauna.

VALUES OF BIODIVERSITY

Biological resources are the basis of life on Earth. Human species are mainly dependent on them for their basic existence. Benefits provided by biodiversity to mankind may be immediate or long term. Biodiversity provides a large range of goods and other services like materialistic, psychological, and spiritual which are vital to our survival. There are mainly two main approaches used to understand and comprehend the value of biodiversity.

1. **Consumptive use valuation:** It involves assessing the direct values of ecosystem functions, soil production, and climatic regulations.
2. **Productive use valuation:** It involves assessing the value of products that are passing through the market. For example, timber, ivory, medicinal plants, fish and meat.

Following are the values, which includes both productive and consumptive uses:

(i) Food includes gram, vegetables, fruits, tea, coffee, condiments, tobacco, liquor oil and so on from plant resources.

(ii) Other uses includes medicines, timber, fuel, fiber, wax, resin, paints, fertilizer, leather and so on.

Some of the astonishing values of biodiversity are:

(a) Initially plants were consumed directly in wild form. But with the course of time, the wild species became the fountain for agriculture. Human beings have used about 5,000 species of plants as food. Out of which only 150 species have entered in the world commerce and about 20 species provide 90 per cent of the world's food.

(b) About three crops, namely, rice, wheat and maize provide 60 per cent of the calories and 55 per cent of the protein.

(c) Fish, cattle, sheep, poultry and even eggs provide the largest source of protein in the world.

(d) Many medicines are made from wild species. In Ayurveda, plant extracts are directly applied or consumed as medicines, while in Allopathic system, 119 chemical substances are obtained from 90 species of higher plants, which are used in medicines. There are about 21,000 plants, which have usage in medicines.

(e) Total world import of cut foliage, plants and cut flowers is equal to US $ 2488 million in 1985.

Values of biodiversity are categorised into social, ethical, aesthetic and optional values.

Social Values of Biodiversity

Biodiversity is valuable as a resource for sustainable development. Biodiversity in terms of social value embraces enjoyment, feeling, imagination, ethics as well as wealth creation which are essential for life. Social values are principles that guide the social behaviour. These values improve standard of living, make people proud and help them to lead happier lives.

Social values of biodiversity through conservation agencies, businessman and government can connect the people and widen the horizon of the society. But, biodiversity does not resonate with the everyday lives of the people. Hence, it does not translate easily into principles to guide corporate and social action. From such backgrounds, opinion makers formulated social values that inspired and motivated the conservation movement and biodiversity.

Ethical Value of Biodiversity

Ethical values are systematic organization of values that establish explicit principles for conduct and behaviour. Ethical values help to maintain both respect for biodiversity and human beings. People are of the view that conservation is detrimental to and compatible with human welfare and economic development arguing that not all the species are necessary. Every species play an important role in maintaining an ecological balance among the living systems of Earth. Loss of many species threatens the survival of many species.

On the other side some people hypothesize that human beings of nature will benefit if we can understand individual components and their functions by studying undisturbed natural ecosystems. Ethical values also stress that man should protect biodiversity because he himself is the main cause for the loss of biodiversity. Mainly the loss of biodiversity is through loss of habitat, exploitations and other perturbations. If this loss of biodiversity is not brought under control, then the lower and middle income-class people are at greatest risk.

Aesthetic Value of Biodiversity

Aesthetic tradition links natural history scientist with the artist and poet, simultaneously. It plays a significant role in professional life of biologists and environmental scientists. Thus, environmental scientists feel compelled to comment on the aesthetic value of biodiversity, as the beauty of biodiversity is inherent and not in the eye of beholder. Biodiversity also provides aesthetic fulfillment for rural people. Aesthetic values of our landscapes and natural systems provide emotional and spiritual well being of a highly urbanized population. According to Orians, "Beautiful landscapes are probably highly functional ones in that they potentially provide rich combination of resources for human existence".

Biodiversity provides fun and recreation. Wild life, pet keeping, gardening are the examples of aesthetic value of biodiversity. Biodiversity is also related to our religious beliefs. For example, tulsi and pipal are considered sacred.

Optional Values of Biodiversity

Optional values are related to the future use of resources and also future flow of information regarding the use of resources. It can be said, optional value reflects the willingness of a risk-averse society to pay a premium, on top of the use-value itself, for guaranteeing access to a resource, the future supply of which is uncertain. Option-values are not immediately seen but have far reaching impact on our living conditions and systems. Few of them are as follows:

(a) Pollination, gene flow etc.
(b) Maintaining water cycles, recharging ground water, penetrating watersheds and buffering for extreme conditions such as flood and drought.
(c) Carbon fixing through photosynthesis which provides support system for species that we harvest.
(d) Soil formation and protection from erosion.
(e) Maintaining nutrient cycles like nitrogen, carbon, sulphur, oxygen and maintaining the oxygen-carbon dioxide ratio.
(f) Decomposing and absorbing pollutants.
(g) Regulating climate at micro and macro levels.
(h) Preserving various values of natural environment.

BIODIVERSITY AT GLOBAL, NATIONAL AND LOCAL LEVEL

Biodiversity at Global Level

At global level there are about five to fifty million species of living forms, out of which only 1.6 million have been identified so far. Out of the above mentioned identified species, 116,8623 species are invertebrates, 49,022 species are vertebrates and protochordata and 3,93,450 species are green plants, fungi, bacteria and viruses. Recently, about 300 new species of fishes have been identified from Amazon region. There are about 18 hot spots all over the world, which cover about 0.74 million sq.km.

Distribution of species is highly uneven. Moist tropical forests occupy about 7 per cent of the land surface and has more than half of the species on Earth. The dominating species are insects and primates.

Reasons for high species diversity

(a) Conditions for evolution are optimum and for extinction fewer in tropical areas.

(b) Species diversity has been conserved over geological time in tropical areas.

(c) Tropical areas are the ideal place for interaction between climate, organisms, time, topography, heredity and parent soil materials.

Biodiversity at National Level

India covers about 32,87,591 sq.km of geographical area and occupies seventh rank as biodiversity rich country. In India about 40 per cent of the total land is under cultivation. India occupies an important place in South Asia due to richness in biodiversity. The biodiversity richness in India is due to variety of climatic, and altitudinal conditions coupled with ecological habitats.

Important Points to Remember

1. India has 10 biogeographical regions.
2. India has two major realms, namely, Palaeoarctic and Indo Malayan and three biomes called the tropical humid forests, the tropical dry forest and warm deserts.
3. India has 13 biosphere reserves, 89 National parks, 27 Tiger reserves, 497 wild life sanctuaries, 5 world heritage and 6 Ramsar wetlands.

Biodiversity at Local Level

Biodiversity at local level means the kind of communities and species present in a particular habitat. It depends on habitat fluxes of energy, nutrient disturbances and organisms of that area.

Biodiversity at local level can be studied under three components:

1. Compositional aspects
2. Structural aspects
3. Functional aspects

1. **Compositional aspects:** It refers to the genetic makeup of population and the type of habitat and communities in an area.
2. **Structural aspects:** It refers to the physical features of a habitat and the horizontal patches of vegetation across a landscape.
3. **Functional aspects:** It refers to the climate, hydrological, geographical, ecological and evolutionary processes that maintain diversity.

INDIA AS A MEGADIVERSITY NATION

India is located between 8°–30° N longitude and 60°–97.5° E altitude with geographical area of 32,87,591 sq.km. There are about 12 countries in the world, which are rich in biodiversity out of which, India is one of the "Megadiversity" nations. Indian megadiversity is due to valuable gene pool upon which the foundation of agriculture, silviculture, pisciculture, horiticulture, piggery, poultry and dairy is built. India has about 45,000 wild species of plants and 81,251 wild species of animals.

The country is lying at the junction of Afro-tropical, Euro-Asian and Indo-Malayan biogeographic realms. India is rich in biodiversity with endemic flora and fauna. The reason for megadiversity nation is due to its favourable environmental conditions. The climatic conditions vary from the humid tropical western ghats to the hot deserts of Rajasthan, from the cold deserts of Ladakh and the icy mountains of Himalayas to the warm coasts of Peninsular India.

Characteristic of India's Megadiversity

India has 10 biogeographical regions, namely, Trans-Himalayan, Himalayan, Deccan Peninsula, Desert, Gangetic plain, Western ghats, Semiarid zone, Coasts, North East India, Lakshadweep Islands and Andaman and Nicobar Island.

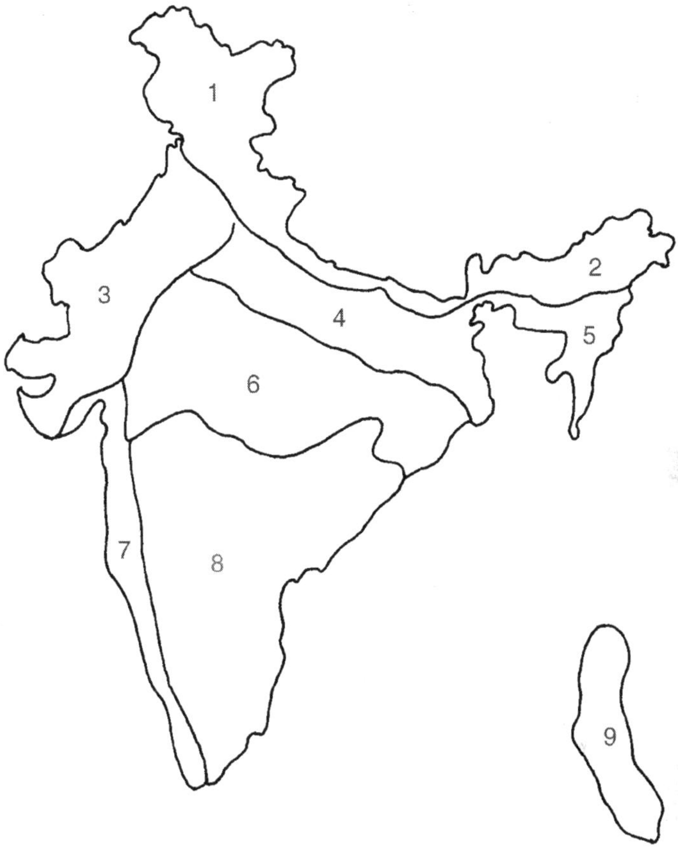

Fig.4.3 Richness of species in India

1. Western Himalayas 2. Eastern Himalayas 3. Indus Plain
4. Gangetic plain 5. Assam 6. Central India
7. Malabar 8. Deccan 9. Andamans

1. It has 497 wild life sanctuaries, 89 national parks, 27 tiger reserves, 13 biosphere reserves, 6 wetlands and 5 world heritage sites.
2. India has two realms namely Palaeoarctic and Indo-Malayan and three biomes namely the tropical humid forests, the tropical deciduous forests and the warm semi-deserts.
3. The country is a Vavilvo's centre of high crop genetic bio-diversity. It is so named after the Russian scientist N.I. Vavilor. It is rich in endemic flora.
4. India is also rich in faunal wealth. It is also rich in marine biota. Marine biota is mainly dominated by sea weeds, fishes, molluscs,

crustaceans, corals, reptiles and mammals. Many species present in the country need identification. The information about flora and fauna is yet incomplete. The western ghats in peninsular India and southern states are rich in biodiversity.

HOT SPOTS OF BIODIVERSITY

Hot spots are those areas which are very sensitive regions as far as biodiversity is considered. These areas are richest and most threatened reservoirs of plant and animal life on Earth. Hot spots can be determined by two criteria. Here we have divided Hot spots in two categories.

Table 4.1 Hot Spots of Indian Biodiversity

Name of the Sensitive area	Location	Reasons for Threat
Andaman and Nicobar island	India	Tourist invasion
Aravali hills	Rajasthan	Habitat destruction
Bharatpur lake and wetland	Rajasthan	Tourist invasion
Chilka lake	Orissa	Fishing
Dal lake	Jammu and Kashmir	Exploitation of aquatic life
Doon valley	Uttaranchal	Deforestation
Gir forest	Gujarat	Population pressure
Himalayas	India and Nepal	population pressure
Kodai lake	Tamil Nadu	Tourist invasion
Little Rann of Kutch	Gujarat	Desertification
Loktak lake	Mahipur	Dryness
Manas	Assam	Population pressure
Ooty lake	Tamil Nadu	Dumping of sewage
Pooyam Kutty valley	Kerala	Construction of dam
Palni hills	Tamil Nadu	Tourist invasion
Pulikat Saltwater lake	Tamil Nadu	Destruction of aquatic life
Narayan Sarovar	Gujarat	Mining activities
North-Eastern Himalayas	Indian subcontinent	Population pressure
Nilgiri hills	Tamil Nadu	Tourist invasion

Megadiversity regions of world

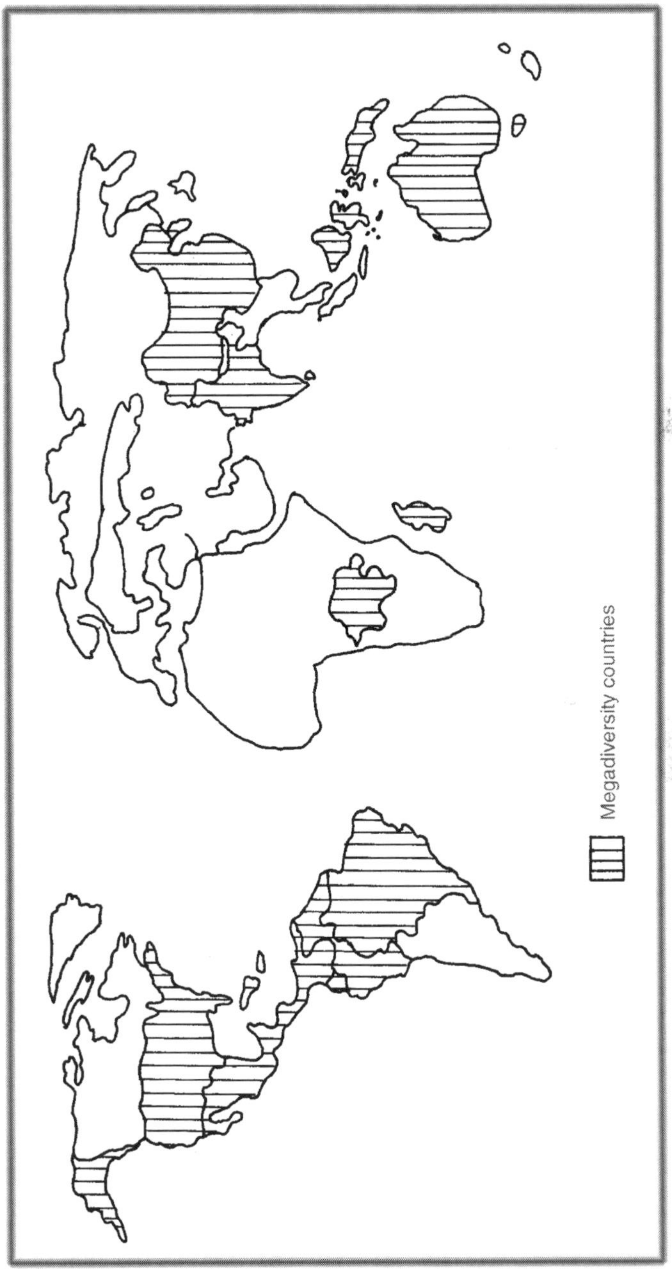

Fig.4.4 Map of world showing biodiversity reserves

Table 4.1 Contd...

Name of the Sensitive area	Location	Reasons for Threat
Renuka lake wetland	Himachal Pradesh	Tourist Invasion
Shivalik hills	Himachal Pradesh	Mining activities
Sariska	Rajasthan	Mining activities
Sunderbans	West Bengal	Erosion of coastal habitats
Thar desert	Rajasthan	Desertification
Western ghats	India	Developmental activities

Table 4.2 Hot Spots of World biodiversity

Name of the Sensitive area	Location	Reasons for Threat
Amazon	Latin America	Deforestation
Arctic Tundra	North pole	Radio Active material
Alaska	USA	Oil Spills
Alps	Europe	Tourist invasion
Antarctica	South pole	Tourist invasion
Brittany seashore	France	Oil spills
Calauit island	Phillipines	Tourist invasion
Caribbean island	South Pacific	Tourist Invasion
Hudson Bay	United States and Canada	Construction of dam
Island of Madagascar	East African Coast	Grazing of flora
Lake Victoria	Kenya	Pollution
Lard Howe Island	-	Exploitation
Maldives Island	South East Asia	Tourist invasion
Masai Mara	Kenya	Tourist invasion
Mauritius	East African Coast	Tourist invasion
River Rhive	Switzerland	Influx of toxic chemicals
St.Helena Island		Over exploitation
Seychelles Island		Over exploitation
Mediterranean Sea Coast		Tourist invasion

THREATS TO BIODIVERSITY

Threat to biodiversity is an important issue in present scenario. On Earth there is no place where there is no ecological catostrophe. Due to which one-third of the species are likely to extinct in the coming years. The tropical forest are shrinking at the rate of 0.8 per cent each year.

Table 4.3 Loss of tropical forests in various ecosystems

Forest	1990 ('000 hectares)	1981–90 (%)
Rain forest	7,13,790	0.6
Moist deciduous	5,91,779	0.9
Hill and Mountain	2,01,417	1.1
Dry Decidous	1,78,579	0.9
Very Dry	59,742	0.5
Desert	8,086	0.9

Causes for Threat to Biodiversity

1. **Climatic change:** Climatic changes are due to global warming, and ozone depletion. Hence these changes result in threat to biodiversity.
2. **Over population:** Due to overpopulation, people encroach more land as a result of which habitat becomes degraded.
3. **Pollution:** Pollution leads to changes in natural habitat. For example, water pollution is dangerous to biotic components of coastal ecosystems.
4. **Introduction of exotic species:** The introduction of exotic species is also a threat to biodiversity because native species are subjected to competition for food and space.
5. **Habitat loss:** Habitat destruction is the main threat towards biodiversity. The loss of wild life habitat is about 80 per cent globally. India has lost about 70 per cent of water ecosystem especially coastal areas. Hence, lots of species have been lost due to habitat destruction. The loss of habitat is due to human activities like mining, industries, dam constructions, and agriculture etc.
6. **Hunting:** From early period, man has hunted for food. But nowadays, people are hunting animals for commercial purpose such as hide and skin, tusk, antlers, meat, pharmaceuticals, cosmetics, fur, decoration and so on. In Africa, 95 per cent of the black rhino have been poached for their horn. The cost of horn is nearly $15,000 in the pharmaceutcal market. Africans have wiped out

about 3000 tonnes of ivory from elephants within seven years. The cost of ivory in the international market is $150 per kg, whereas in India the cost varies from Rs. 2,000 to 3,000 per kg. The major buyers are Hong Kong and Japan.

Several species of spotted cats, scarlet macaw have been eliminated from the range of Central America. 70,000 whales were slaughtered in 1962. Therefore, international ban is put on the whale products. In India, tigers are hunted for bones and skin, musk deer for musk, crocodile for their skin, elephants for ivory, rhino for horns, whale for whale bone or baleen. Baleen is used to make combs and so on.

Indian desert lizard is hunted to extract oil. This oil has aphrodisiac properties.

7. **Over exploitation:** Over exploitation not only leads to loss of economic species but also biological species such as insectivorous species for laboratory work. Examples are Psilotum, Ghetum and Nepeuthes. Besides biological species, medicinal and horticulture plants are also under threat due to over exploitation. Excessive harvesting of marine animals such as whales, molluscus, sea turtles, sea cows results in extinction of the species.
8. **Collection for zoo and research purpose:** Flora and Fauna are collected for zoos, laboratories and research purpose in university and institutes to know the anatomical, physiological and genetic similarities to human beings.

Table 4.4 Forest area loss due to human activities

Activity	Area lost (in Mha)
Transmission lines and roads	0.061
Industries of townships	0.134
River-valley projects	0.502
Agriculture	2.623
Miscellaneous	1.008
Total area lost	4.328

Source: The state of forest report (1987) Ministry of Environment and Forests.

9. **Control of pests and predators:** Pests and predators are killed by control measures.
10. **Deforestation:** Deforestation is one of the main causes for the loss of wild life. In India, deforestation rate is about 13,000 sq km annually that results in a huge loss of species everyday as shown in table.

Poaching of Wildlife

Poaching of wildlife is the illegal killing of charismatic megaverte-brates. Poachers use a variety of methods to obtain duty free vension. Popular method of poaching is the use of rifle. Second method is the use of cross bow. Third method of poaching is sharing. But this method is not popular. It occurs globally due to several reasons.

1. The wild animals are killed for their products like antlers, fur, meat, hide, skin, pharmaceuticals, perfumes and cosmetics for decoration purposes.
2. Many species of spotted cats have been jeopardized for the demand of their fur.
3. Indian tiger is poached for tiger bones, rhinos for horns, bear for their body parts, musk deer for musk, elephant for ivory, gharial and crocodile for their skin and spiny tailed lizard for oil.
4. 95 per cent of the black rhino population has been exterminated by poachers for their horn in Africa.
5. 1/3 of the African elephants have been wiped out for 3000 tonnes of ivory.
6. Poaching of Indian tiger has risen because of tiger bones for pharmaceutical purposes.

Thus poaching results in threat to all wild life resources and depletion of the collective heritage of all citizens who respect and comply with legislation and regulations. Illegal trade and poaching of African tuskers has been reduced due to CITES regulations. Carving and sale of African ivory has been banned in India from April 2, 1992.

Man–Wildlife Conflict

Man and wildlife conflict is manifest between local communities and the wild life protecting bodies. The protection of wild life areas is an indirect conflict with the economic interests of local communities. The rural communities have not only lost their wildlife resources, but they also bear the full costs of crop damages caused by migrating wild life. Rising rural populations and poverty, has further deteriorated the situation. Hunting, harvesting and illegal encroachment are the only available means of securing income. Local populations are not directly involved in these activities, but their alienation from wild life resources, forces them to oppose the exploitation of these resources by others.

The breakdown of traditional common property management regimes into virtual open access exploitation leaves rural communities with little means to enforce sustainable management. Several programs have been attempted to counteract this by diverting some of the revenue generated by sustainable management of wild life populations into the hands of the communities in which these populations are concentrated.

An example of man-wild life conflict is human-elephant conflict. The elephant commands the affection of the people. Elephants are intelligent, lovable. For the majority of the human inhabitants, they are the creatures to be admired and cherished. However, once you put some of these elephants next to your house, where they could not only rob you, your food, your livelihood and possibly your life, attitude swiftly changes to conflict.

Endemic and Endangered Species of India

Endemic species: Those which are confined to a particular area and do not extend beyond one region, that is, a country, mountain range or an Island.

Table 4.5 Diversity of wild plants/organisms in India

Flora	India	% of Endemism
Angiosperms	15,000	
Gymnosperms	64	60%
Pteridophytes	1022	N.A.
Bryophytes	2584	N.A.
Algae	2500	N.A.
Fungi	23000	N.A.
Bacteria	850	N.A.
Lichens	1600	N.A.
N.A.: Not available		

Source: Biodiversity in India, R.R. Rao NBRI Lucknow 1994.

Features of Endemic species

1. They are restricted to a particular area.
2. These species are adapted to a particular environment.
3. Endemic species have little variability due to poor mutability.
4. Endemic species result due to lack of exchange of gene pools.

Endangered species: Endangered species are in danger of extinction and their survival is unlikely if the casual factors of decline continue operating. It is denoted by 'E'. The list of endangered species was compiled in 1970 in Red Data Book.

Endangered species of wetlands: *Nymphaea tetragona, Utricularia minor, Nelumbo nucifera, Limnopoa meeboldii, Uydrocera triflora, Hydrilla polysperma, Euryale ferox, Eriocaulan hunile, Callitriche verna, and Aldrovanda vesiculosa.*

Endangered species of Thar desert: About 10 per cent of the existing biological diversity is endangered. The flora of Thar desert includes tumba, phog, babool, bordi, kumat, kikar, kejri, guggal, ker, sewan, nagauri ashgandh, rohida.

Endangered species of plants: *Commiphora wightii, Tecomella undulata, Eleocorpus recurvatus, calligonum, Polygonoides, Lasianthus ciliatus, Viola pilosa, Aspagragus fysonil, Vanda caerulea, Areca concinna, Euphorbia abdelburi, and Renanthera.*

Endemic Species	
Animals	**No. of species**
Protozoa	
Lepidoptera	9
Free living	87
Parasitic	532
Mollusca	
Land and fresh water	807
Pisces	
Fresh water	64
Marine	14
Amphibia	85
Reptilia	152
Aves	50
Mammalia	37
Total endemic animal species	1837
Plants	
Steridophyta	200
Angio sperms	4,950
Total endemic plant species	5,`50

Source: Arnon 1997, National action plan on biodiversity, Ministry of Environment of India, New Delhi

CONSERVATION OF BIODIVERSITY

Conservation of biodiversity is defined as the rational use of the environment to provide high quality of life for human beings.

Aims of conservation

1. To preserve biological diversity
2. To maintain ecological processes and life support systems.
3. To ensure sustainable utilization of species and ecosystem.

For the social and economic development of a country conservation is very important. Earlier, biodiversity conservation was very much limited for saving genes, species and habitats. But now-a-days, conservation of biodiversity is based on sustainability and equitability of the welfare of society.

Steps for Conservation of Biodiversity

1. International biodiversity strategy programme organized by World Resource Institute.
2. Funding for conservation in developing countries.
3. International financial support for biodiversity conservation and identification are on priority.
4. Transfer of technologies for the conservation of biodiversity.
5. World conservation strategy which involves the link between living resource conservation and sustainable development.

Methods for conservation

There are mainly two methods of conservation of biodiversity.

In situ conservation

Ex situ conservation

In situ conservation: This involves the conservation of genetic resources. It mainly includes national parks, nature reserves, sancturies, biosphere reserves, cultural landscapes and so on.

Ex situ conservation: This type of conservation is outside their habitats such as germplasm for seeds, ova, cells, pollen, gene pools and so on. Plants are easily conserved than animals. This includes seed banks, pollen storage, tissue culture, botanical gardens.

Institutions involved in conservation of biological diversity fall under the ministry of environment, agriculture, science and technology.

Our late Prime Minister Smt. Indira Gandhi stressed on conservation and worship of trees and animals in 1980. She said, "The interest in conservation is not sentimental one but the rediscovery of a truth well known to our sagas. The Indian tradition teaches us that all forms of life, human, animal and plants are so closely linked that disturbance in one gives rise to imbalance in the others."

REVIEW QUESTIONS

Short Type Questions

1. Define biodiversity.
2. What are the different levels of biodiversity?
3. What are values of biodiversity?
4. What are social values of biodiversity?
5. What do you mean by ethical value of biodiversity?
6. What is the aesthetic value of biodiversity?
7. How many national parks and sanctuaries are there in India?
8. Name the richest hot spots in India.
9. What are the major causes of extinction of species?
10. What are the hot spots of biodiversity?
11. What kind of threats to biodiversity leads to the extinction of species?
12. What do you mean by endemic species?
13. What is species diversity?
14. What is ecosystem diversity?
15. What are IUCN and WWF?
16. Mention different types of conservation of biodiversity.
17. What are endemic species?
18. What are endangered species?

Descriptive Questions

19. Explain different biogeographical areas in India with respect to diversity.
20. Explain the optional values of biodiversity.
21. What are the reasons for richness of biodiversity in India?
22. Why is India known as 'Megadiversity' nation?
23. Discuss in detail the threats to biodiversity.
24. Explain the poaching of wild life.
25. Discuss the man-wild life conflicts.
26. Discuss in detail endemic and endangered species of India.
27. Discuss in detail the conservation strategies of biodiversity.

5

Environmental Pollution

Chapter Objectives

- Definition
- Types of Pollution
- Disaster Management

DEFINITION

Any change in the physical, chemical and biological characteristics of air, water and soil that affect the health or activities of human beings or other forms of life in an undesirable manner is the result of pollution. Pollution not only causes physical harm but also psychological distress, foul odours, unpleasant sights, noise, heat etc. In other words, pollution is a direct or indirect change in any component of the biosphere that is harmful to the living components.

Pollution is caused by pollutants. Some of the main pollutants are gases, metals, deposited matter, poison, sewage water, fertilisers, radioactive substances, heat and noise. These pollutants are of two types–biodegradable pollutants and non-biodegradable pollutants. The substances that decompose slowly in the natural environment are known as non biodegradable whereas those substances that can decompose naturally in the environment are known as biodegradable, such as sewage waste. The sewage wastes are rapidly decomposed by natural processes. Non biodegradable substances accumulate along the food chain and became biologically magnified.

TYPES OF POLLUTION

Pollution is of two types:

Natural pollution: Pollution that originates from natural processes.

Artificial pollution: Pollution that originates from the activities of man.

Air Pollution

When the concentration of normal component of air or some chemical added, or formed in the air upto the point of causing harm to the living organisms or to materials. The addition of chemical to air leads to pollution of air is called pollution or atmospheric pollution. Air pollution is one of the most dangerous pollution and reported mainly in the metropolitans and industrial area. The first case of air pollution was found in Belgium 1990.

Sources of air pollution A major source of air pollution results from the gaseous emissions from automobiles, industries, or domestic combustion, burning of fossil fuels and so on. Thousands of air pollutants are present in the troposphere. Some of the pollutants that cause air pollution are as follows:

Manmade sources of air pollution: These are the sources of air pollution that result from the activities of man.

1. *Carbon oxides:* Carbon monoxides and carbon dioxides come from automobiles and industries.
2. *Sulphur oxides:* Sulphur trioxides, Sulphur oxide comes from thermal power stations and industries.
3. *Nitrogen oxides:* Nitrogen oxide, nitrogen dioxide and nitrous oxide are pollutants from industries.
4. *Volatile organic compounds:* These are Benzene, methane and CFCs. These compounds are emitted from refrigerators, air conditioners, thermal power stations and industries.
5. *Photochemical oxidants:* Hydrogen peroxides, aldehydes, hydroxyl radicals, PANs, and ozone. These pollutants come from automobiles and industries.
6. *Suspended particulate matter:* It includes soot, dust, lead, arsenic, nitrate, sulphates sulfates and salts, oil, dioxins, pesticides. All these pollutants come from automobiles, industries, metallurigical processes and thermal power stations.
7. *Radioactive substances:* These include iodine–131, Plutonium–239, strontium–90 and Radon–222 produced from nuclear reactions into the atmosphere in the form of suspended particulate matter and gases.
8. *Noise:* This pollution comes from aeroplanes, buses, trains, autorickshaws, machines, radios, and so on.
9. *Heat:* When fossil fuels are burnt, they give rise to heat.

Natural sources of air pollution: Natural sources of air pollution are forest fires, pollen dispersal, wind erosion of soil, volcanic eruptions, bacterial decomposition, and radioactive substances located in the earth.

Indoor sources of air pollution: Radon-222 gas, asbestos, and smoke from cigarette causes most dangerous indoor air pollution. Air inside the car, buildings, home and schools also emit severe air pollution. The indoor air pollutants cause coughing, sneezing, headaches and burning of eyes. In developing country like India cooking fuels like dung-cakes, coal, wood which produce carbon dioxide, carbon monoxide and sulphur dioxide are used. Burning of cigratte issues benzene and pyrene which causes cancer and tuberculosis.

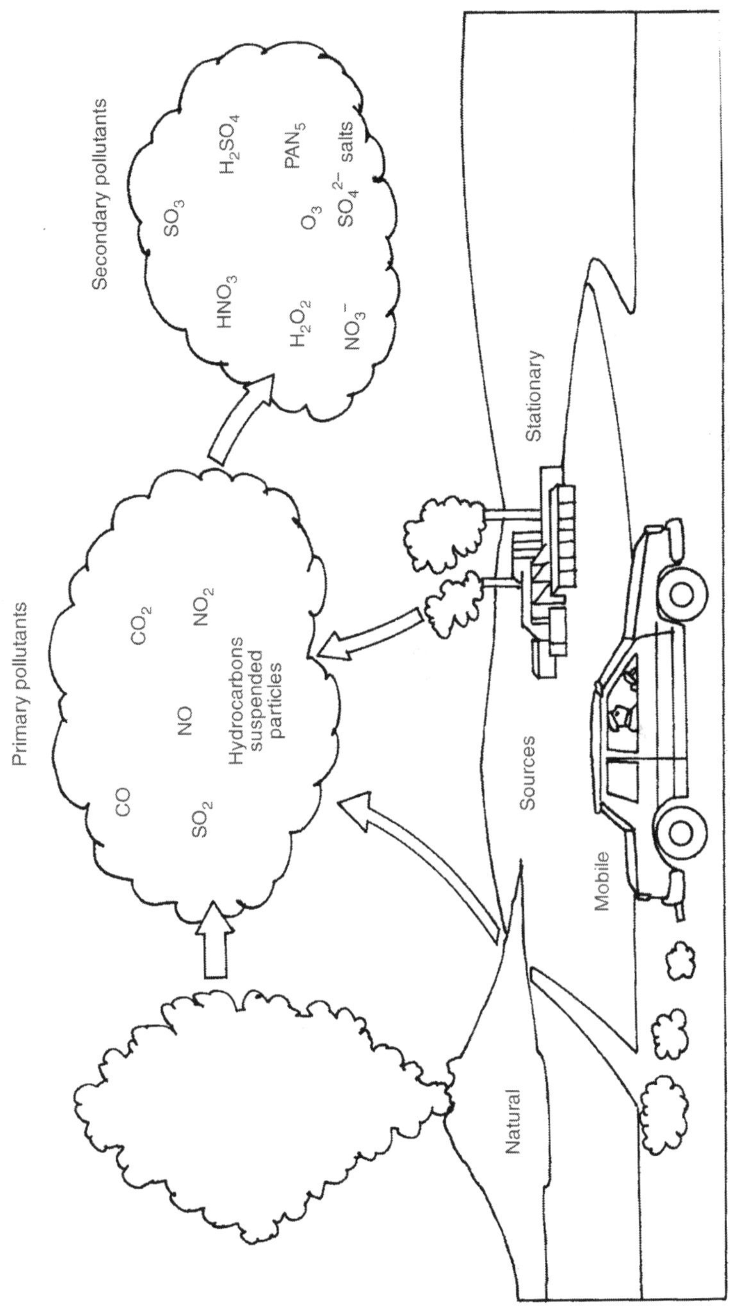

Fig.5.1 Primary and Secondary Pollutants

Table 5.1 Common air pollutants, their sources and health effects on man (after Southwick 1976)

S.No.	Pollutants	Sources	Effects on Human Beings
1.	Ammonia	Dye making, explosives, fertilizers	passages. Inflame upper respiratory
2.	Aldehyde	Thermal decomposition of oil and fats	Irritate respiratory passages
3.	Arsenic	Processes of arsenic soldering	Red cells of blood get reduce, kidney failure
4.	Chlorines	Bleaching of chemical processes	Attack on mucous membrane of eyes, respiratory diseases
5.	Carbon monoxides	Burning of coal, exhausts	Reduces oxygen carrying capacity of blood
6.	Hydrogen sulphide	Bituminous coal, refinereis, industries	Rotten eggs smell, eyes and throat irritates
7.	Hydrogen fluorides	Refineries, fertilisers, aluminium production	Irritating, disrupts body passages
8.	Hydrogen cyanides	Furnaces, metal plating, chemical manufacturing	Nerve cells get damaged, throat becomes dry, vision not clear
9.	Nitrogen oxides	Exhausts	Dust penetrate into the lungs
10.	Phosgenes	Dye manufacturing	Irritation and coughing
11.	Sulphur dioxides	Oil and coal combustion	Headache, vomiting due to chest constriction
12.	Suspended particles	Incinerators	Cancer and eye irritation

Table 5.2 Indoor Air Pollutants

S.No.	Pollutants	Source	Effects
1.	Chloroform	Chlorine treated water in hot showers	Cancer
2.	Paradichloro	Air freshners, moth ball	Cancer
3.	Tetrachloro ethylene	Dry cleaning–fluid fumes on clothes	Nerve disorders damage to liver and kidneys
4.	Formaldehyde	Furniture stuffing paneling, particle board, foam insulation	Irritation of eyes, throat, skin, lungs, nausea, dizziness

Table 5.2 Contd...

S.No.	Pollutants	Source	Effects
5.	Benzo-a-pyrene	Tobacco smoke, wood stoves	Lung cancer
6.	Styrene	Carpets, plastic products	Kidney and liver damage
7.	Radon–222	Radioactive soil and rock surrounding foundation, water supply	Lung cancer
8.	Methylene chloride	Paint strippers and thinners	Nerve disorders, diabetes
9.	Carbon monoxide	Faulty furnaces, unwanted gas stoves, kerosene heaters, wood stoves	Headaches, irregular heartbeat, drowsiness
10.	Asbestos	Pipe insulation, vinyl ceiling and floor tiles	Lung Cancer
11.	Nitrogen oxides	Unvented gas stoves and kerosene heaters, wood stoves	Cold, headaches and unvented gas stoves
12.	1,1,1–Tri-chloroethane	Aerosol sprays	Dizziness irregular breathing

Effects of air pollutants

Following are the effects of air pollution:

1. **Green house effect:** When there is an increase in carbon dioxide concentration in the air, the thick layer of carbon dioxide gas does not allows the heat to be re-radiated out, and return back to Earth. Thick carbon dioxide layer functions like the glass panels of a green house which allow the sunlight to filter but prevent the heat to re-radiated into the outer space. This is called **greenhouse effect**. It results in increase in carbon dioxide concentration in the air and thereby causes an increase in the temperature of the surroundings. This increased temperature in the atmosphere is called **global warming**.
2. **Effect on human health:** Air pollutants cause a number of respiratory diseases such as asthma, bronchitis, and lung cancer. Sulphur dioxide causes constriction of respiratory passages and

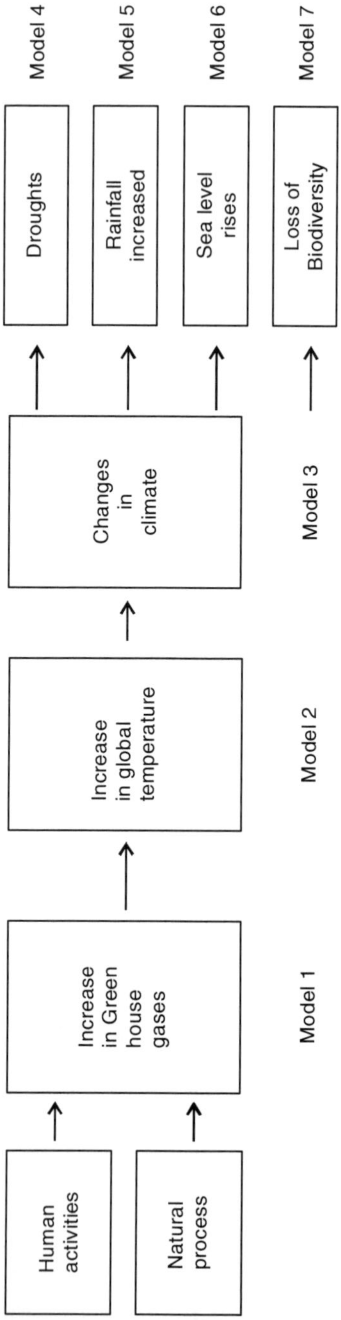

Fig. 5.2 Model for Green house effect

results in bronchitis. Suspended particulate and sulphur dioxide enters into lungs (dusting) asthma. Carbon monoxide causes dizziness and unconsciousness. Inhaling of ozone causes coughing, throat irritation, shortness of breath and heart troubles. Smoking can also cause emphysema. It is an incurable disease that reduces the ability of the lungs to transfer oxygen to the blood. Emphysema also reduces the elasticity of lungs.

3. **Effect on plants:** Air pollutants break down the waxy coating of the leaves that prevents excessive water loss. Conifer trees, spruce and fir are at higher elevations and highly vulnerable to the effects of air pollutants. Air pollutants also inhibit the photosynthesis and plant growth. Sulphur dioxide, ozone, nitrogen oxides and PAN are the main pollutants that damage leaves of crop plants and trees. Acid deposition enters the soil and kills microorganisms in soil,

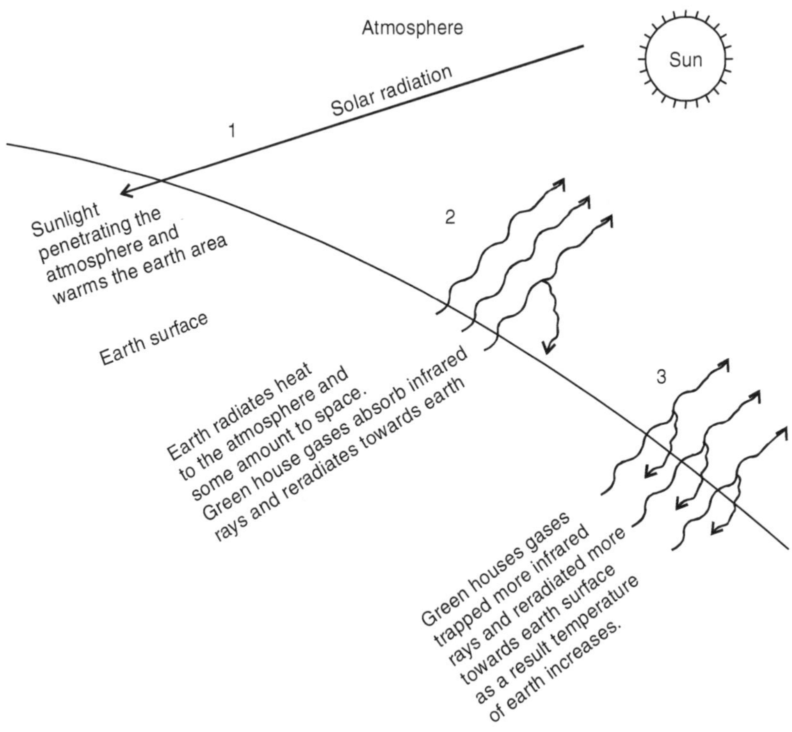

Fig.5.3 Green house effect (The unity and diversity of life 5th ed. Belmont, Calif. Wadsworth 1989)

damages root filaments and reduce uptake of water and nutrients. Air pollutants also make trees vulnerable to frost, insects, drought and diseases. Some of the pollutants make soil so acidic that new seedlings do not survive at all.

4. **Effect on animals:** There is little information about the health effects of polluted air on animals. A cattle grazing on grass contaminated with fluoride and acid deposition causes reduction in milk production, weakens their bones causing florosis and ultimately leading to death.
5. **Effects on aquatic life:** Acid deposition over the fresh water lakes has negative impact on aquatic life, particularly fishes. In Sweden sensitive lakes contain no fish because of acidity of water.

Table 5.3 Effects on materials

S.No.	Materials	Effect	Air Pollutants
1.	Fabrics	Fading and damage	Ozone, sulphur dioxide, nitrogen dioxide
2.	Rubber	Cracks	Ozone
3.	Leather	Cracks	Sulfur dioxide
4.	Metals	Corrosion	Hydrogen sulphide, nitric acid, sulphuric acid
5.	Paints	Decolourisation	Ozone, sulphur dioxide, hydrogen sulphide
6.	Paper	Decolourisation	Sulphur dioxide

Historic buildings, stained glass windows, irreplaceable marble statues are also decoloured by the air pollutants.

Control of air pollution: Air pollution can be reduced mainly by two approaches. First is the input approach and second being the output approach. Input approach reduces the pollution whereas output approach treats the symptoms. First approach is easier and cheaper in the longer run than the second approach.

Following are the methods of control of air pollution:

1. Reduce, reuse and recycle the waste of paper, metals etc.
2. Regulate population growth
3. Reduce energy use
4. Use solar energy, wind energy and flowing water

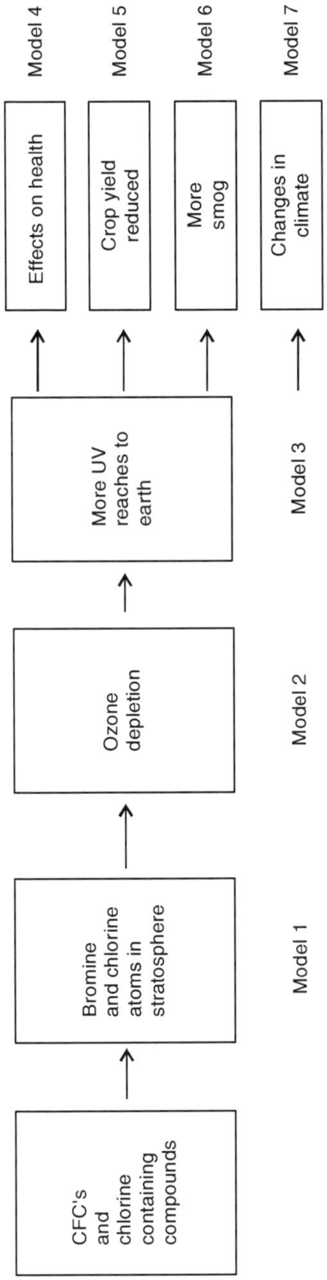

Fig.5.4 Model of ozone layer depletion

5. Burn low-sulphur coal
6. Remove pollutants after combustion
7. Use tall smoke stacks in industrial plants
8. Improve fuel efficiency of motor vehicles
9. Use less polluting fuels in vehicles
10. Rely more on para transit or mass transit
11. Car, scooter and other vehicles should be in good condition
12. Emphasis on stage of less-polluting automobiles engines
13. Shift to less-polluting fuel
14. Use emission control devices to work effectively
15. Reduce indoor air pollutants by using house plant
16. Do not use air freshners
17. Install air-to-air heat exchangers

Water Pollution

Water pollution is any biological, physical or chemical change in surface water or ground water that can adversely affect the living organisms. It is a aquatic contamination which eliminates living organisms from a rivers, lakes, ponds or ocean.

Sources of water pollution Water pollution is of two types:

Point and Non-point Sources

1. **Point sources:** Point sources are those that discharge pollutants, through ditches, sewers, pipes into water bodies at specific areas, coal mines, oil tanks, electric power plants, industrial and municipal point sources.
2. **Non-point Sources:** Non-point sources are those that discharge pollutants in large areas. For example, runoff into surface water, seepage into ground water from logged forests, urban and suburban lands, parking lots, roadways, etc. The main sources of water pollution are sewage, industrial effluents, agricultural discharges, industrial wastes, fossil fuel plant, and nuclear power plant.

Types of Water Pollution

1. Ground water pollution
2. Soil water pollution
3. Surface water pollution
4. River pollution

5. Lake pollution
6. Coastal water pollution
7. Estuarine pollution
8. Open-ocean pollution
9. Bacterial pollution
10. Viral pollution
11. Metallic pollution
12. Nutrient pollution
13. Petrochemical pollution
14. Pesticide pollution
15. Radioactive pollution
16. Thermal pollution

Sources of water pollution

1. **Domestic sewage waste:** It is a water borne waste that comes from habitations animals and food processing plants. It involves paper, cloth, human and animal excreta and detergents etc.
2. **Industrial effluents:** Inorganic and organic pollutants are present in effluents. Wastes comes from tanneries, dying textiles, paper mills, breweries, steel industries, and mining areas etc. The main pollutants are grave, oil, plastics, heavy metals, phenols, toxins, acids, dyes, salts, plasticizers, cyanides and DDT etc.
3. **Agricultural discharges:** It involves chemicals such as fertilizers, insecticides, pesticides, herbicides, artificial fertilisers, biocides etc.
4. **Radioactive substances:** It involves radioactive waste that comes from nuclear power plants.
5. **Minerals:** Minerals comes from mining belts and areas.
6. **Sediments:** These results from land erosions.
7. **Detergents:** These come from washing, bathing and also from manufacture of detergents.

Agricultural and industrial pollutants are discharged into the world's oceans annually (after Southwich 1976). Refer Table 5.4

Effects of water pollution: Water pollution affects the abiotic and biotic factors of different aquatic bodies and its effects on man remains drastic. Some of the effects of water pollution are as follows:

1. **Production of disease causing agents:** Production of bacterial, viruses, protozoa and parasites tranmits typhoid fever, cholera, infections, hepatitis and such similar diseases. These pathogens enter the environment through faeces and urine of infected organisms.
2. **Oxygen demanding wastes:** Domestic sewage, manure, biodegradable organic waste deplete the dissolved oxygen of water. Depletion in oxygen content leads to release of phosphates and causes eutrophication in lakes.

Table 5.4

S.No.	Pollutants	Source	Annual discharge (metric tons)
1.	Mercury	Industries	100,000
2.	DDT	Agriculture and public health wastes	25,000
3.	Benzene hexachloride	Agriculture and public health wastes	50,000
4.	Air borne lead	Motor vehicles	350,000
5.	Air borne hydrocarbons	Motor vehicles, power plants and industries	15,000,000
6.	Aldrin-Texaphene	Agriculture and public health	25,000
7.	Petroleum and industrial hydrocarbons	Oil tankers, industrial wastes, offshore wells	3,405,000
8.	Polychlorinated biphenols	Plastic industries	25,000

3. **Inorganic plant nutrients:** Addition of inorganic nutrients such as nitrates and phosphates results in growth of algae and other plants which when die consume oxygen from water for decomposition. Death and decay of plant results in changes in concentration of carbon dioxide that changes pH, of water bodies.
4. **Toxic compounds:** Toxic compounds involve heavy metals, biocides, fungicides, pesticides, cyanides and other organic and inorganic compounds that are harmful to aquatic life. The toxic compounds are non-biodegradable in nature that shows bio magnification at various levels of food chain. DDT, aldrin and dieldrin are toxic compounds that are banned. These show biomagnification in the food chain due to non-biodegradable nature and deposition in fatty tissues.
5. **Radioactive substances:** These substances are amplified to higher concentrations in the food chains. These substances cause birth defects, genetic damage and cancer.

Table 5.5 Effects of heavy metal water pollutants on human beings

S.No.	Metal	Effects
1.	Zinc	Kidney failure, vomiting
2.	Copper	Coma, fever, hypertension
3.	Cadmium	Bone weakening, kidney failure, hypertension, nervous system damage, reduced growth
4.	Lead	Damage of liver, kidney failure, vomiting, loss of apetite
5.	Mercury	Pain of chest, diarrhoea, abdomen pain, headache
6.	Arsenic	Liver damage, lung cancer, kidney failure, ulcer, mental disorder

6. **Heat:** Excessive input of heat makes aquatic organisms more prone to diseases and toxins. The temperature of water body increase whereas oxygen content decreases with the rise in temperature.
7. **Sediments:** Insoluble soil particles destroy spawning grounds of fishes, reservoirs and stream channels. These substances also makes water cloudy and break down the food webs.

Control of water pollution Control of water pollution is very important otherwise the entire ecosystem becomes unbalanced. Following are the few methods by which we can prevent water pollution.

1. Establish nationwide standards for water contaminants.
2. Total ban on non-biodegradable substances.
3. Controlled use of pesticides and fertilizers.
4. Sewage waste should be properly discharged.
5. Industrial waste should be properly treated before discharge into water bodies.
6. Nitrates and phosphates should be removed by treatment to prevent eutrophication.
7. Chlorination should be done before disinfection of drinking water.
8. Recycling is the best method of prevention and control of water pollution.

Marine Pollution

Oceans have vast areas and have the ability to consume unlimited waste products. But there are few reports to indicate the amount of pollutants being dumped into the oceans and coastal areas.

There is an African Proverb, "Water may flow in a thousand channels, but it all returns to the sea". Baltic and Mediterranean seas are vulnerable to water pollution because their coasts are highly populated and industrialized.

Sources of marine water pollution

Following are the sources of water pollution:

1. Natural run off
2. Garbage and untreated sewage from ships
3. Agricultural and urban runoff
4. Atmospheric fallout
5. Accidental oil spills from tankers and offshore oil drilling platforms
6. Barges and ships dump industrial wastes
7. Dumping of radioactive wastes
8. Dumping of waste heat
9. Lubricating oil from industries
10. Metal industries
11. Paint industries
12. Removal of ballast water

Effects of marine water pollution

1. Pollution affects sensitive flora and fauna such as phytoplankton, zooplankton, coral reefs, algal species, fishes, invertebrates, birds and mammals.
2. Magnification of pollutants affect fishes, waterbirds and other marine biota.
3. Shrinkage of coral reefs in tropical areas.
4. Reduced productivity of pelagic organisms.

Control of marine water pollution

The marine water pollution can be minimised by following methods.

1. Effluents and other wastes should not be discharged into rivers and oceans.
2. Sewage waste should be properly treated in sewage treatment plant.
3. Industrial effluents should be treated before discharge.
4. Dumping of radioactive substances should be avoided.

5. Developmental activities of coastal areas should be minimised.
6. Dumping of toxic wastes, sewage sludge should be banned.
7. Oil ballast should not be dumped into the sea.
8. Overflow of sewer should be prevented.
9. Sensitive coastal areas should be protected by not allowing drilling.
10. The best method of prevention and control of marine water pollution is the recycling of different wastes.
11. Strictly regulate safety, training and operation procedures for offshore wells.
12. Enforcement of marine acts and coastal acts.

Soil Pollution

Soil is the topmost layer of the earth that is produced by weathering of rocks. Soils are the best homes for a variety of living organisms. But simultaneously soils are more vulnerable to pollution from various sources. Soil pollution is caused by solid wastes and chemicals. Major problem in the cities is disposal of solid wastes such as crop residues, animal manure, and industrial wastes like fly ash, cinders, chemicals. These are the residues of combustion of garbage, paper, plastic, rubber, leather, brick, sand, metal, cardboard and glass, dead animals like dogs, cats, birds, discarded products such as washing machines, scooters, T.V., refrigerators, and other gadgets.

Sources of Soil Pollution

Following are the sources of soil pollution:

1. Chemicals emitted into the air such as radioactive substances, sulphur, minerals and lead ultimately come down to surface to pollute soil.
2. Herbicides and pesticides are applied on the crop plants. But they ultimately reach to the soil and remain in the soil for a longer period of time.
3. Soil of the Earth have huge amount of animal and human excreta that contains many pathogenic organisms.
4. Industrial wastes has biodegradable and non-biodegradable substances such as metals, toxic substances, paint, cloth ray, plastic, heavy metals etc.

5. Domestic wastes are also dumped into the soil.
6. Sewage wastes and sludge in the urban areas are major sources of soil pollution.
7. Explosion of radioactive substances and wastes comes from industries, laboratory, etc. into the soil. Main radioactive substances are isotopes of uranium, thorium, radium, caesium, strontium and radium etc. These radioactive substances remain in the soil for a longer period and emit radiations.

Effects of soil pollution

Following are the effects of soil pollution:

1. In forest ecosystem pollutants inhibit process of soil formation.
2. Pesticides and herbicides cause a disruption in species composition of communities.
3. Some of the non-biodegradable chemicals persist into the soil and affect soil flora and fauna.
4. Sewage sludge causes many diseases.
5. Soil pollutants change physical, chemical and biological properties of soil.
6. It reduces soil fertility due to toxic chemicals which are drained as industrial effluent.
7. Radioactive isotopes fall on the trees, herbs and shrubs and enter the food chain of grazing animals and human beings that result in abnormalities such as, instead of calcium, strontium gets deposited in the bones. The bones become weak due to deposition of strontium.
8. Toxic chemicals contaminate soil that gets leached down into the ground water and ultimately causes diseases to human beings and animals.
9. Fertilisers containing phosphorus and nitrogen enter into the water bodies near the agricultural fields and cause eutrophication.

Control of Soil Pollution

We can minimise the soil pollution in the following ways:

1. For solid waste disposal, controlled tipping or sanitary landfill method is adopted in Delhi. In sanitary landfill, two metres of refuse is covered by 23 cm of Earth up to the level of the hole chosen. The surface is then used for housing.

2. Effluents should be treated before discharge into the agricultural fields and nearby areas.
3. Use of inorganic fertiliser should be in the controlled manner.
4. Solid wastes should be properly disposed off.
5. Recycling of solid wastes should be emphasised.
6. Cattle dung should be used for biogas plants particularly in the remote areas and villages.
7. Sewage wastes and sludge should be properly recycled and treated.
8. Recycle and reuse the liquid and solid waste.
9. Biodegradable substances should be processed for microbial degradation for reducing the soil pollution.

Thermal Pollution

Industries utilise cold water for various processes and the resulting hot water flows into lakes and streams that cause thermal pollution. Thermal pollution has harmful effects on the aquatic flora and fauna. Coal fired generators and atomic energy plants produce a large amount of waste heat which is carried away as hot water.

Sources of thermal pollution

Following are the sources of thermal pollution:

1. Removal of vegetation from the banks of streams and lakes cause thermal pollution
2. Thermal power plants
3. Nuclear power plants
4. Refineries, steel mills
5. Power plants
6. Heat producing industries
7. Various industrial processes that use cold water and discharge hot water.

Effects of thermal pollution

Following are the effects of thermal pollution.

1. Hot water reduces the dissolved oxygen that results in death and decay of water plants and animals.
2. Hot water also increases the respiration rate of aquatic organisms

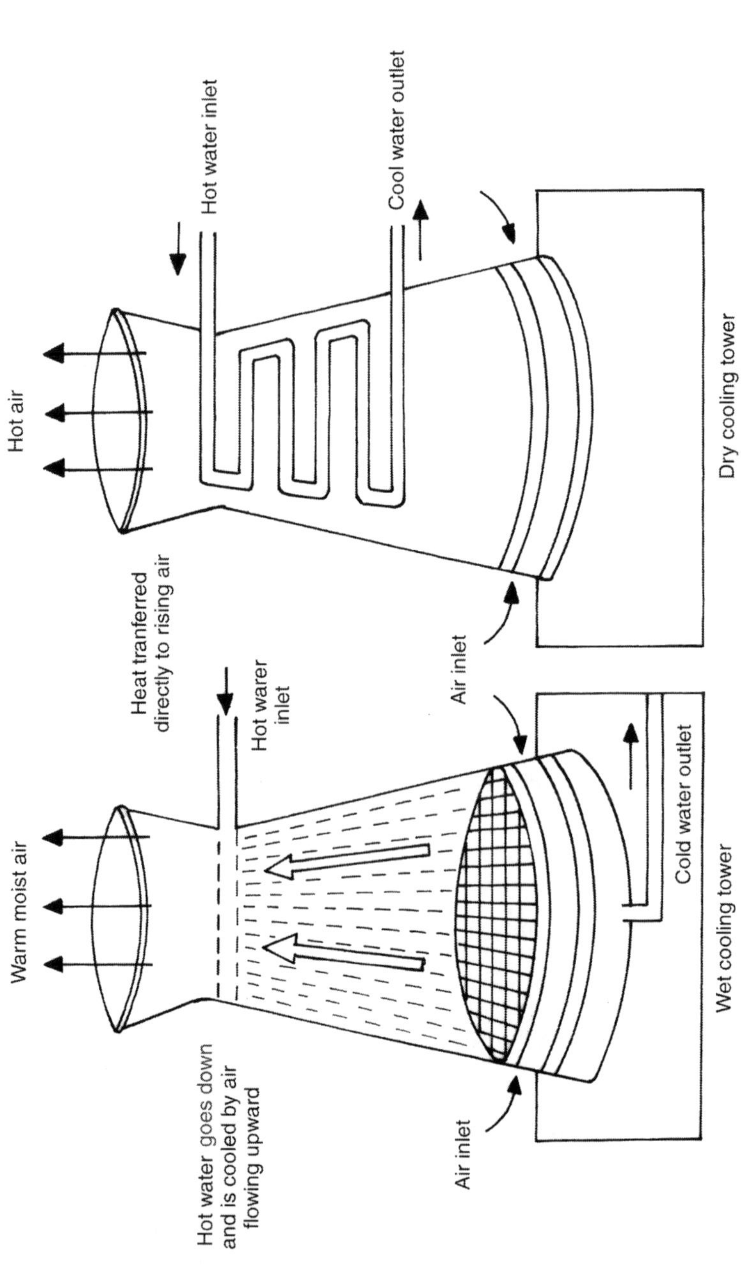

Fig.5.5 Cooling towers transfer heat to the cooling water and into the atmosphere

that increases their susceptibility to diseases, toxic chemicals and parasites.

3. Hot water also disturbs spawning of fishes and kills them.
4. Hot water lengthens the commercial fishing season.
5. Hot water reduces the winter ice cover in cold areas.
6. Rise in temperature of water increases rate of chemical reactions that results in decay of the organic matter, and leads rusting of iron.
7. Hot water also increases rate of exchange of salts in organisms and thus affects the organisms.
8. Oxygen penetration reduces deep water due to high temperatures.

Control of thermal pollution

Following are the methods to minimise thermal pollution:

1. Discharges of hot water should be at a point away from the vulnerable shore zone.
2. Transfer the heat from the water to the atmosphere as shown in figure 5.5 wet and dry cooling towers.
3. Hot water should be discharged into shallow ponds and allow it to cooled for reuse.
4. The most common method is to withdraw cold water from a water body and pass it through a plant then return the heated water to the same water body.
5. Hot water discharged should be into the same water body.
6. Cooling towers are installed for recycling the hot water and cools water then discharged into water bodies.
7. Spray towers are installed. These towers reduces the heating effect of water.

Noise Pollution

A loud, unwanted or unpleasant sound that causes discomfort is called noise. The release of unwanted sound in the environment is called noise pollution. This is most widespread occupational hazard. Noise pollution is mainly caused due to technological advancement. To measure levels of noise, sound pressure is measured in decibel. This is denoted by db. High pitched sounds are more louder than the low pitched sounds. Sound pressure damage at 75 db or higher.

Sources of noise pollution

1. Pollution from transportation.
2. Construction equipments
3. Vaccum cleaners
4. Sirens
5. Unwanted loud music
6. Earphones
7. Entertaining equipments
8. Air conditioners
9. Blasting
10. Use of crackers
11. Newspaper press
12. Rocket engines
13. Electrical home appliances
14. Industrial operations
15. Celebrations

Effects of noise pollution

1. Noise increases the level of cholesterol in the blood.
2. Noise causes headache and increases blood pressure.
3. Noise causes the pupils of the eye to dilate and mental abnormalities in foetus.
4. Noise pollution affects the hearing ability of human beings.
5. Noise pollution causes anxiety and stress.
6. Noise pollution causes hormonal imbalance.
7. Noise detracts attention and cause emotional imbalance.
8. Noise interferes with man's communication and disturbs concentration.
9. Noise pollution causes physical and psychological damage to human beings.
10. Noise pollution damages heart, liver and brain.

Table 5.6 Different sounds at different noise levels

S.No.	Different Sounds	Noise levels (dB)
1.	Military rifle	150
2.	Air raid siren	140
3.	Earphones	130
4.	Stereo in car	120
5.	Rock music	110
6.	Chain saw	100
7.	Lawn mower	90
8.	Factory	80
9.	Vacuum cleaner	70
10.	Normal conversation	60
11.	Rainfall	50
12.	Quiet room	40
13.	Rural area	30
14.	Whisper	20
15.	Normal breathing	10

Table 5.7 Noise levels (dB) CPCB recommended noise standards of different zones

	Zones	Night	Day
1.	Silence zone	40	50
2.	Residential zone	45	55
3.	Commercial zone	55	65
4.	Industrial	70	75

Table 5.9 Noise levels of some cities across the world

S.No.	Cities	Noise levels (dB)
1.	Kathmandu	75
2.	Delhi	80
3.	Mumbai	82
4.	Calcutta	85
5.	New York	88
6.	Rome	99
7.	Nanjing (China)	105

Control of noise pollution

Following are the methods to minimise the noise pollution:

1. Government should enact laws to penalise noise pollution at various social functions. Blowing of horn should be minimised in vehicle congested areas.
2. Plantation should be done with broad leaves especially conifers trees which absorb noise.
3. Industrial areas, railway stations and aerodromes should be away from residential areas.
4. Proper oiling should be done in machines to reduce noise pollution.
5. Cotton plugs or ear puffs should be used by the occupational workers.
6. Sources of noise pollution should be reduced such as old vehicles, heavy vehicles and so on.
7. Different types of sound absorbing media should be used for noise making machines.
8. Sound absorbing silencers should be used for minimise the noise such as fibrous materials.

Nuclear Hazards or Radioactive Pollution

Nuclear hazards are mainly radioactive substances. These radioactive substances are present in nature. These substances undergo a natural decay and give rise to alpha and beta gamma particles. These are fast moving particles with high energy radiations.

Radioisotopes emit radiations that produce ionising radiations. The main ionising energy released from radioisotopes are gamma rays. Gamma rays are a form of high energy electromagnetic radia-tion. Particles with high speed are emitted from the nuclei. They are alpha particles and beta particles. The penetrating power of alpha, beta and gamma particles is different. Alpha particles are fast moving particles with positive charge.

Beta particles are high speed with relative more penetration than alpha particles. Gamma radiations are more penetrating than beta and alpha radiations.

Many radioactive isotopes are found in rocks and soils in natural form such as uranium–235, radium–226, thorium–232, potassium–40, carbon–14. Others are cobalt, iron, krypton, plutonium, strontium. The radioactive isotopes have serious environmental implications. The isotopes involved are Cobalt–60, Cesium–137, Iodine–131, Krypton–85, Strontium–90, Plutonium–239, Argon–41, Strontium–90 has a half-life of 24 years and behaves like calcium in the nutrient cycles. Thus it is absorbed by plants and in turn by animals and gets accumulated in the bone tissues.

Radioactive phosphate, iodine–132 and cesium accumulate in plants and animals through food chain.

Sources of radioactive substances

Mainly there are two types of sources:

1. **Natural sources:** It involves cosmic rays from space and terrestrial radiation from radio-nucleosides present in Earth's crust, soil, air, water, food and rocks.
2. **Man-made sources:** It involves mining, refining of platinum and thorium nuclear power plants, test laboratories, nuclear explosions, X-rays, diagnostic, bits and radiation therapy.

Effects of radio-active substances

Following are the effects of radioactive substances:

1. Exposure to radioactive substances causes leukemia and bone tumors.
2. It causes genetic damage and infant mortality.
3. Radioactive substances get deposited in the bone tissue through food chain and causes serious health hazards to man.
4. Ionising radiations effects burns, dead tissues and death of organisms.
5. Ionising radiations cause incidents of tumors, cancer and shortening of life span.
6. Non-ionising radiations cause injury to the skin.
7. Non-ionising radiations cause sunburns.
8. Non-ionising radiations damage eyes during solar eclipses.
9. Non-ionising radiations affects the cells and low levels of multicellular organisms.

Control of radiations

Nuclear hazards can be minimised in the following ways:

1. Installation of nuclear power plants should be carefully done.
2. Nuclear accidents should be prevented.

3. Use of nuclear weapons should be banned.
4. Proper disposal of radioactive wastes.
5. Leakage of radioactive elements from laboratories and reactors should be totally checked.

Solid Waste Management

Solid waste is useless, unwanted or discarded material. United States generates about 10 billion metric tons of solid waste. The main reason for solid waste production is the high levels of population and consumption of resources.

Table 5.11 Sources of Solid Waste generated in United States

	Sources of Solid Waste	Solid Waste Produced in Percentage
1.	Mining and Oil and Gas Production	75 per cent
2.	Agricuture	13 per cent
3.	Industry	9.5 per cent
4.	Municipal	1.5 per cent
5.	Sewage Sludge	1 per cent

(Data from environmental protection agency and U.S. Bureau of mines)

The largest solid waste generated is from mining and agricultural activities. The second largest category is of solid wastes such as rocks, sand, slag and so on. Hence, if waste generation is continuous, it creates a problem. Thus, it is important to manage and minimise the solid wastes.

Sources of solid wastes

Following are the sources of solid wastes:

1. **Wastes from industries:** It includes packing material, plastic, paper, fly ash, rubber, acids, sludge, batteries, asbestos, metals, organic wastes and slag etc.
2. **Wastes from mining:** It involves abandoned and unrestored metal, stones, gravel pits and abandoned sand.
3. **Wastes from municipal:** It involves wastes from habitations, urban areas, paper, paper board, metals, plastic, backyard waste, food, glass and other materials.
4. **Waste from parks and gardens:** It includes dead, dry leaves and some residue and so on.

5. **Waste from slaughter houses:** It includes remains of slaughtered animals.
6. **Waste from nuclear power plants:** It involves radioactive substances.
7. **Waste from thermal power plant:** It include fly ash in solid form.
8. **Waste from building construction:** It includes wood, concrete, cement, soil particles, rubber and marble etc.
9. **Waste from market and shopping mall:** It include packing material, paper, plastic, vegetable waste etc.
10. **Waste from hospital:** It includes syringes, pathological waste, anatomical waste, glass bottles, polythene bags and scrap metal etc.
11. **Litter as a waste:** It is also a source of waste. For example, helium filled balloons in parties often burst and become litter.

Effects of solid waste

Following are the effects of solid waste:

1. Changes in physico-chemical and biological properties of soil and plants.
2. Environment becomes unhygenic due to heaps of solid waste.
3. Production of open breeding sites of insects and infectous organisms.
4. Heaps of solid waste can cause cancer by burning of carcinogenic waste.
5. Ground water gets contaminated.
6. Air becomes polluted due to foul smell.
7. Soil becomes polluted due to which crop yield becomes poor.

Management of solid wastes

Mainly there are two ways to deal with the management of solid waste. First is waste management and second is waste prevention.

Waste prevention can be done in two steps.

First step involves following measures:

1. Change in industrial processes, so as to eliminate the use of harmful chemicals.
2. Purchase of different products.
3. Less use of a harmful product.
4. Reduce packaging mateirals in products.

5. Make products that last longer and are easy to repair, recycle and reuse.

Second step involves:

1. Repair products
2. Reuse products
3. Recycle products
4. Use of compost products
5. Buy reusable and recyclable products

Waste management is the second method for solid wastes that involves following steps:

1. Treatment of waste to reduce toxicity
2. To incinerate waste
3. Burn the wastes in land fills
4. Release waste into environment for dispersal

Vermiculture and composting are the two techniques used for waste management. Vermiculture is a technique that converts solid wastes into compost by the action of earthworms. Composting is a technique that converts organic waste into compost and carbon dioxide by degradation process. Microorganisms help in the degradation of organic wastes. The urban solid wastes can be converted into manure by means of anaerobic digestion process. The compost is a fertilizer for a soil and it is also used as animal feed. This approach is used in many European countries such as Netherlands, France, Germany, Italy and Sweden.

Case Study of Pollution in Virginia

American Viscose Corporation built a rayon plant outside the town in 1938. This rayon plant provided 3,000 Jobs for people of nearby area. Whenever wind blew, the air smelled like rotten eggs, the smell of hydrogen sulphide and carbon disulphide gas. The paint of the houses converted into black. But in 1940 no one gave attention because the rayon plant provided boost to the local economy. But after 1980, environmental officials charged the company of polluting the nearby ground water and Shenandoah River of South fork. Shenandoah River is the main water source for Washington. The Environmental Protection Agency (EPA) in 1988 discussed ways of cleaning up the rayon plant. It would cost about millions of dollars and take atleast 10 to 30 years. Hence the company filed for bankruptcy.

Case Study of Rayon Plant in Virginia

In 1938 American Viscose Corporation built the world's largest rayon plant outside the town. The plant provided employment to about 3000 people. But when the plant was in working order it emited toxic hydrogen sulphide and carbon disulphide gas that gives rotten eggs smell and blackened fumes. But within a few days the surrounding buildings were blackened due to fumes.

In 1980 environmental officials charged the company because this artex fibres was discharging industrial effluent in the Shenandoah river and nearby ground water. This Shenandoah river is a major water source in Washington.

In 1984 Environmental Protection Agency (EPA) declared the sites as the highest level of hazardous wastes in the country and put a penalty of 9.3 $ million for toxic wastes water cleanup.

In 1988 State Attorney General again put penalty of $ 19.7 million for ground water pollution.

Cleanup of river and plant costs million of dollars and takes a decade. As a result the company filed for bankruptcy.

Case Study–Poland most polluted country

Poland was declared as the most polluted country in 1985 by Polish Academy of Sciences. Water, air and soil of Poland country is fully contaminated with toxic wastes due to industrial and coal burning power plants. People of Poland country suffer from respiratory illnesses, cancers and so on. Vehicles turn on their head lights during the day. 78 per cent of forests showed damage due to pollution. Poland cities do not have any waste water treatment plants. They dump the wastes into vistula river, which empties into the Baltic Sea. 40 per cent of the water is unfit for industrial use and 96 per cent of the water is not monitored for drinking after disinfection. Soil is contaminated with heavy metal and unfit for crop production.

Poland is rapidly implementing modern pollution control equipments and energy efficiency technologies.

Role of Individual in Prevention of Pollution

Role of individual in prevention of pollution is very important for a healthy environment. It is the duty of every citizen to protect and improve the natural environment. Each individual should change his life style in such a manner so as to reduce pollution. These are following ways or roles by which individual can prevent the environmental pollution.

1. Environmental awareness program should be conducted for modification of attitude of remote area people towards the nature. We should educate them by providing knowledge of plants, animals, microorganisms and their relationship with the environment.
2. Educate every section of society irrespective of sex and caste to protect and improve the environment.
3. Use hazardous chemicals only when absolutely necessary and in the small amounts.
4. Use less hazardous and cheaper cleaning products such as baking soda, vinegar and borax for cleaning and bleaching of clothes. Baking soda can be used as tooth paste and deodordant.
5. Do not flush hazardous chemicals down the toilet.
6. Reduce use of fossil fuels by using mass transit, walking or bicycling.
7. Use energy efficient light bulbs and other home apliances.
8. Use solar energy and natural gas for heating space and water.
9. Cool your home by planting trees.
10. Plant more trees for reduction of carbon dioxide and cool the environment.
11. Buy only those products that are made from recycled materials.
12. Take steps to reduce the population growth rate.
13. Do not use the products containing chloro and fluoro carbons, carbon tetrachloride.
14. Avoid using aerosol spray products because these products emit hydrocarbons or other propellant chemicals.
15. Use refrigerators and freezes that use helium as a coolant.
16. Car and home conditioners should be checked regularly to avoid CFC's leakage.
17. When building a home, save as many trees as possible.
18. Landscape the surrounding area of the habitations.
19. Use minimum amount of water.
20. Use biogas plant for kitchen purposes.
21. Use eco-friendly products for day to day life.
22. Always follow three R's – Recycle, Reuse and Reduce the wastes.
23. Use natural pesticides instead of synthetic pesticides.
24. Impose heavy penaltiies for motor vehicles exceeding emission levels of pollution.

25. Enforce "non-smoking" in public places.
26. Support soil, water and air pollution prevention laws and their enforcement at the local, state and national levels with emphasis on pollution control.

Disaster Management

Disasters are the natural or man induced untolerable catastrophic losses of property and lives. The disaster includes cyclones, floods, earthquakes, volcanic eruptions and droughts. The intensity of a disaster is always viewed in terms of the damage done to the human society. Earthquakes and volcanoes are due to internal Earth processes endogenetic forces whereas floods are the result of external Earth processes (exogentic forces).

Floods

The massive runoff is known as flooding. Flooding is caused by heavy rain or rapid melting of snow. Flooding is one of the common types of natural disaster similar to volcanic eruptions and earthquakes. Flooding kills thousands of people and causes tens of billions of dollars of damage every year. In China, in 1959 about two million people were killed due to floods followed by famine and diseases. Analysis revealed that flooding causes about 39 per cent deaths, cyclones 36 per cent and earthquakes 13 per cent. Flooding not only damages the property but also causes of greater economic loss. It also causes contamination in flooded area. The main cause of flooding is the removal of vegetation.

The vegetation can be removed by overgrazing, construction of buildings, forest fires, mining activities, timber extractions and urbanisation. Urbanisation is also responsible for flooding by replacing vegetation and soil with highways that lead to massive runoff of rain water.

Flooding can be reduced or managed by following methods:

1. **Reforestation:** Maximum number of plants should be grown so that runoff water is retained by the roots of plants and soil.
2. **Flood frequency curve:** This curve will give an idea about the frequency of floods based on past history.
3. **Rechannelization:** Channelisation deepens and widens a section of stream. Thus the stream can accomodate higher discharge of water.

4. **Flood control dam:** Dams controls flood water which can be used for irrigation and recreational facilities.
5. **Artificial leaves:** These are built where there are no natural ones. These are in the form of sandbags.
6. **Flood plain management:** In this, a beautiful green belt is built covering several kilometres. Such examples are seen in Arizona and Scottsdale.
7. **Construction of ponds:** Ponds should be constructed in urban areas to hold rain water.
8. **Diversion of rain water:** Rain water should be diverted through storm sewers.
9. **Restoration of wet lands:** Wet lands should be restored for holding maximum water.
10. **Water absorbing vegetation:** Grow more water absorbing vegetation such as eculyptus.
11. **Overgrazing:** Overgrazing should be under control.
12. **Vegetation:** Vegetation should be grown along with highways, shopping centre, parking lots and office building etc.
13. **Mining:** Mining should be avoided in flood prone areas.

Earthquakes

Earthquake is caused due to the sudden movements of Earth's crust. The Earth's crust is made up of many plates of rocks that keep on moving slowly and along their boundries. Due to friction these plates are prevented from slipping as a result stress is build up, which results in sudden fractures within the plates along the boundaries and causes earthquakes. The intensity of an earthquake can be measured by Richter scale devised by Charles F. Richter in 1935.

When the Richter scale is less than four then the Earthquake is insignificant. When the richter scale is on 4 to 4.9, then it is called a minor. When richter scale is within 7 to 7.9 then the earthquake intensity is of major type and when more than 8 then it is destructive in nature.

The place of the origin of an earthquake inside earth is called hypocentre and on the surface of earth, it is known as epicentre. Earthquakes are caused due to disequilibrium in any part of Earth's crust. The largest earthquake occured on May 22, 1963 in Chile with the magnitude of 9.5 on Richter scale. In India a major earthquake hit Bhuj town in Gujarat. Recently, on 26 December 2004, Tsunami created by earthquake hit Kerala, West Bengal, Andaman, Pondicherry, Tamil Nadu and Andhra Pradesh in India, Indonesia, Thailand, Malaysia, Sri Lanka and Maldives. In China in 1963 about 500,000 people were killed due to earthquake. Coastal areas are also affected by earthquake generated water waves known as Tsunamis. Tsunamis travel at a speed of 950 kilometres per hour with increase in speed towards seashore.

Human activities are also responsible for increasing earthquake magnitude.

1. Underground nuclear testing has caused minor earthquakes.
2. Deep-well disposal of liquid waste also gave rise to earthquakes.
3. Addition of more water in reservoirs behind dams also increases earthquake activity. For example, Lake Mead behind Hoover Dam.

Impacts of earthquake

1. Earthquake sets up shock waves that radiates out from the centre of the movement known as Focus. Focus is the point of initial movement of earthquake and epicenter is the point on the surface directly above the focus. Hence, when stressed part of Earth's fracture energy is released in the form of shock waves and it radiates outward from the focus point.
2. Earthquake causes flooding in the low lying area.
3. Earthquake causes Tsunami waves near coastal areas.
4. Liquefaction of earth's crust causes buildings to sink.
5. Landslides may occur on hilly grounds.
6. Tremendous loss of life and property.
7. Various types of mass wasting.

Preventive measures for earthquake activity are as follows:

1. Wooden houses should be preferred in earthquake prone areas.
2. Dams should not be constructed in earthquake prone areas.
3. Identification of terrain risk areas by means of remote sensing and electronic techniques.
4. Declare earthquake prone area by historic information from local people of that area.

Landslides

Landslides occur when rocks of soil masses move down due to the gravitational pull. Chemical action of water causes weathering of rocks making it prone to landslides. Landslides include shaking and permanent vertical or horizontal displacement of the ground. Sudden landslides are more dangerous than the slow landslides. Water and vegetation are mainly responsible for landslides. Water causes weathering of rocks due to its chemical activity. That makes the rock more prone to landslides. Similarly vegetation also increases the activity of landslides. The plant material consolidates the soil materials and provides cohesion by its root system that retards the flow of water. This increases the landslides activity. Landslides are governed by the forces that tend to pull earth material down. Land slides are also triggered by volcanic eruptions and earthquake activity.

Landslides can be marked by many factors such as:

1. Saturation with water of the unconsolidated sediments.
2. Unconsolidated sediments exposed by house building.
3. Disturbances in resistant rock overlying the rock of low resistance.
4. Volcanic eruptions and earthquakes.

Control of landslides is a difficult task. However, these can be minimised by certain methods.

1. Provide wired stone blocks to support slope
2. Support at the base of a slope by concrete.
3. Draining the subsurface and surface water.

Landslides results in great loss of life and property.

Cyclones

Cyclones occur in tropical coastal areas due to heat and moisture in the warm oceans. These are giant circular storm systems that developes due to pressure gradient. These storms develop in late summers or early fall. These cyclones cover about 320 kilometers per hour. Cyclones bring formential rains and accomplished by storms and squarks. Cyclones occur when the surface temperature of the sea is above 26°C. Tropical cyclones move like spinning top with a speed of 10 to 30 kilometer per hour. Cyclones originate over oceans and derive energy from the latent heat of evaporation of water sucked into low pressure centre. Their paths are rough and have high tidal surges. The devastation by cyclones is more when storm surge coincide with the astronomical tide. Sea water rushes in lands and inundates the low lying area by the combined force. Cyclones are named as hurricanes in the Atlantic, Willy Willies in the sea around Australia, Typhoons in the western pacific, carribbean in Northeastern

Pacific and cyclones in the Indian ocean. More cyclones occur in the Bay of Bengal than in the Arabian Sea. About six to seven cyclones occur in the year out of which four to five are severe.

Management of cyclones: is very difficult but some measures can be taken to minimise the devastation.

1. Grow more trees in the coastal areas.
2. Proper drainage system should be developed.
3. Roads should be wide for quick evacuation.
4. Construction of dams.
5. Storm shelter.
6. Wind breaks.
7. Predictions should be made by remote sensing and electronic techniques.

REVIEW QUESTIONS

Short Type Questions

1. Define pollution.
2. Name different types of pollution.
3. Define pollutants.
4. Name two types of air pollutants.
5. What is air pollution?
6. What is water pollution?
7. Define soil pollution.
8. Define thermal pollution.
9. What is noise pollution?
10. What are nuclear hazards?
11. What is solid waste management?
12. Define floods.
13. What is an earthquake?
14. What are cyclones?
15. What are landslides?

Descriptive Questions

1. What are the sources of air pollution?
2. Discuss the effects of air pollution.
3. Describe the phenomenon green house effect and global warming.
4. Explain the control of air pollution.
5. What are the sources of water pollution?
6. What are the effects of water pollution?

7. How can water pollution be controlled?
8. Discuss sources, effects and control of marine water pollution.
9. Explain sources, effects and control of soil pollution.
10. What are the sources, effects and control of thermal pollution?
11. What are the sources, effects and control of noise pollution?
12. What do you know about ocean water pollution?
13. What are radiation? Discuss their effects on human health.
14. How solid wastes can be managed?
15. What are the effects of solid wastes towards environment?
16. What is the role of individual in prevention of pollution?
17. How floods can be managed?
18. What are the factors responsible for earth quakes?
19. How landslides can be controlled?
20. Discuss in detail the minimization of cyclones.

6

Social Issues and the Environment

CHAPTER OBJECTIVES

- Unsustainable to Sustainable Development
- Urban Problems Related to Energy
- Water Conservation
- Resettlement and Rehabilitation of People: Its Problems and Concerned Case Studies
- Environmental Ethics

UNSUSTAINABLE TO SUSTAINABLE DEVELOPMENT

All underdeveloped countries are facing a critical situation on economic and environmental fronts. For maintaining economic growth they have to give more emphasis on agriculture and industries but at the cost of environment. Depletion of resources sets in environmental degradation, which in turn weakens the economy of the nation. Second major issue of underdeveloped countries is population explosion. If population growth continues in the same manner, we will soon face doomsday. This will inevitably lead to unsustainable development and eventually to collapse of the interrelated system of Earth.

Sustainable development leads to the fulfilment of the present needs without compromising with the ability of future generations to meet their own needs. The term 'sustainable development' emerged in Earth Summit held at Rio de Janerio, Brazil in 1992. This term came in agreement Agenda-21, which proposed a global programme of action for sustainable development in social, political and economic context for the 21st century.

Main Features of sustainable development

There are mainly two features which are discussed below.

Intra-generational equity It refers to the development processes that seeks to minimize the wealth gaps between nations. It mainly emphasizes that the benefits of technology should seek to achieve the goals of intra-generational equity. The technology development for the underdeveloped countries like producing drought-resistant varieties of seeds, and vaccines to prevent infections and diseases will support the economic growth. This will also reduce the wealth gap and lead to sustainable development.

Inter-generational equity It refers to the optimum utilization of resources and environment for future generations. We should hand over a healthy, safe and resourceful environment to our future generations.

Achievement of Sustainable Development

Sustainable development can be achieved through the following means:

- **Population stabilization:** We should take steps to stabilize our population. If population is more, then the consumption of resources

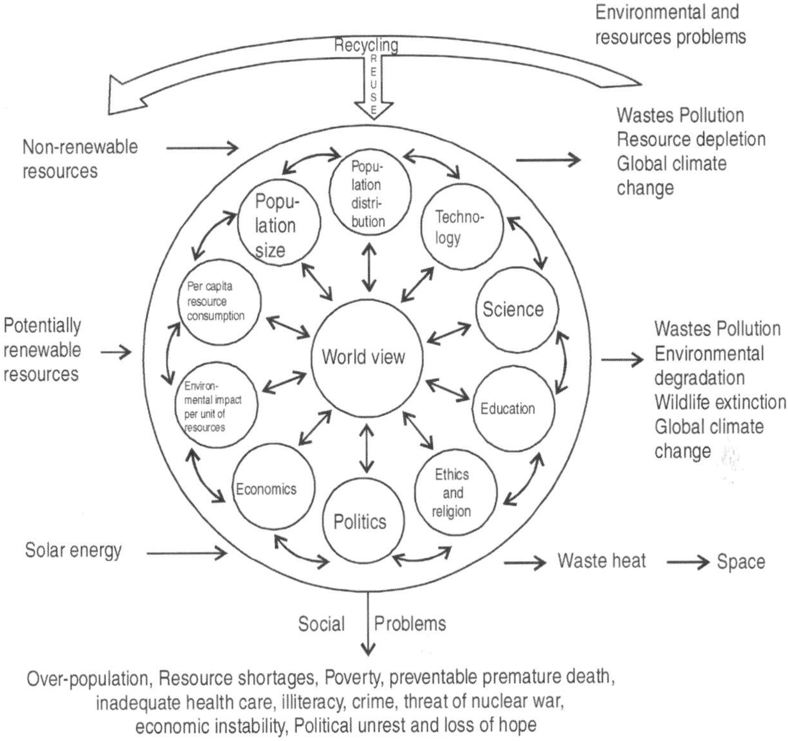

Fig. 6.1 Simplified model showing resources, environmental and social problems are due to complex interacting factors

will be more; as a result it, will cross the carrying capacity and lead to environmental degradation. This will lead to unsustainability.

- **Land use planning:** There should be proper use of land by laying emphasis on polyculture and cultivation of perennial crop.
- **Minimum use of energy:** There should be minimal use of non-renewable fossil fuel energy to reduce pollution and dependence on expensive oil.
- **Heavy reliance on human labour:** It is one of the most abundant resources in the developing countries.
- **Conservation of soil:** There should be emphasis on conservation of soil, soil nutrients and water to increase short and long term crop productivity and reduce water pollution.
- **Control of pollution:** There should be emphasis on control of air and water pollution to improve air quality and proper use of water for irrigation.

- **Three R's:** Recycling, reuse and reduction of wastes and residues in order to achieve sustainability.
- **Environmental awareness and education:** It will help in changing the thinking of the people towards the environment.
- **Proper use of technology:** The technology used should be eco-friendly, resource-efficient and produce minimum waste.

URBAN PROBLEMS RELATED TO ENERGY

Energy plays an important role in the socio-economic development of society. Energy requirement in urban areas is more as compared to rural areas. Urban areas are the centres for trade, education and employment. Due to these activities majority of rural people shift to urban areas. As a result urban growth is fast. Hence due to unplanned and uncontrollable growth, a lot of waste is generated and lot of energy is consumed. Following are urban problems related to energy as mentioned below:

- Industries use a lot of energy.
- Transportation uses a lot of fuel.
- Electrical gadgets like washing machine, mixers, ovens, geysers and so on utilise maximum energy.
- Residential lighting.
- Water supply is the biggest problem in urban areas as a lot of electricity is utilised for supply.
- Multiplexes, buildings of schools, colleges and shopping malls also require a lot of energy in the form of light.
- Air and water pollution is controlled by energy dependent technology.
- Waste generation is properly disposed off by electrical energy based technology.

Hence, in urban areas energy requirement is very huge due to dense population and high waste generation as compared to rural areas.

WATER CONSERVATION

Water conservation is very important because 30 per cent to 50 per cent of the water used is unnecessarily wasted. The world is heading towards a water crisis. West Asia and Africa are worst affected by water scarcity.

Methods for Conservation of Water

1. **Reducing irrigation losses**
    - It can be reduced by trickle or drip irrigation.
    - By growing hybrid crop varieties which use less amount of water.
    - Irrigation efficiency can be improved by using computer-controlled systems.
    - Irrigation ditches can be lined with plastic to prevent seepage.
    - Ponds can be constructed to store runoff.
2. **Reducing evaporation losses**
    - It can be reduced by using conservation tillage covering ponds and covering the soil with mulch.
3. **Use of less water for industry and domestic purposes**
    - There should be recycling of waste water.
    - Use treated water for irrigation.
    - Use grey water for washing like cars, pavements, watering kitchen garden and so on (Grey water is washwater except toilet wastes and food wastes derived from garbage grindees).
    - Store water in soil root zone in humid regions.
    - Reduce runoff losses which can be done by terrace farming, contour cultivation and water storage system.

Watershed Management

The land area that delivers water, sediment and dissolved substances via stream to a major river and ultimately to the sea is called watershed and drainage basin respectively. The management of runoff is called watershed management. It involves proper utilisation of water resources for maximum production and causes the minimum damage to the natural resources. Himalayas is one of the most important watersheds in the world.

Methods of Watershed Management

- **Afforestation:** Afforestation reduces the runoff and loss of fertile soil.
- **Agroforestry:** Crop plantation along with trees reduces the soil erosion and helps in retention of soil moisture.

- **Water harvesting:** Storage of water should be done properly for dry seasons with the help of dams, tanks etc.
- **Bunding:** This prevents runoff discharges and soil loss in Shivaliks.
- **Strip cropping:** This prevents soil loss from crop fields.
- **Terracing:** This also prevents soil erosion.
- **Mining:** By improper mining, soil get disturbed and results in landslides and erosion.
- **Overgrazing:** Overgrazing should be prevented to check soil erosion and landslides.
- **Thin wiring:** On sleeper slopes with sliding faces, straw, tied with thin wire helps in establishing vegetation and stabilizing the slopes.

Rain Water Harvesting

Rain waster harvesting means capturing of rainwater and storing it underground. The aim of rainwater harvesting is to check the runoff water. There are varieties of method of harvesting water.

Methods of rain water harvesting

- Capturing runoff from roof tops and storing in tanks or reservoirs below the ground.
- Capturing runoff from local catchments.
- Capturing seasonal floodwaters from local streams.
- By constructing pits, lagoons.
- By constructing dams etc.
- By recharging the ground water.

Aims of rain harvesting

Aims of rainwater harvesting are as follows:

- Avoid flooding.
- Meet increasing demands of water.
- Reduce runoff loss.
- Reduce groundwater contamination.
- Raise water table by recharge ground water.

Techniques for rain water harvesting

1. **Modern Techniques**
    - By building underground water tanks, evaporation loss is minimum and there are reduced chances of contamination of water.

- Roof top rainwater harvesting is done by diverting the rain water to surface tank or pit which can be later used for different purposes.

2. **Traditional Techniques**
   - Roof top rain water is stored into storage tank.
   - Diversion of rain water in Telaabs, Johars, Hauz and Baawaris etc.

RESETTLEMENT AND REHABILITATION OF PEOPLE: ITS PROBLEMS AND CONCERNED CASE STUDIES

Rehabilitation and resettlement of people in the plains causes socio-economic problems. Following are the main problems and concerns faced by the displaced people.

- Displacement of native people causes economic and psychological distress.
- It deprives them of forest products like vegetables, fruits, fodder, fuel wood, tuber and roots.
- It deprives them of drinking and irrigation water from natural springs and streams.
- They lose the grandeur of the forest ecosystem, and methods of recreation and enjoyment.
- Tribal people are conditioned to hilly or forest environment. It becomes difficult for them to settle in plains and open environment.
- Displaced people lose their ancestral rights on the forests.
- Loss of close link between the people and environment.
- Break up of families occurs.

Factors Responsible for Rehabilitation and Resettlement

Construction of dams The construction of dams and river valley projects is the most serious reason for the rehabilitation. India is one of the countries in world leading to big dam constructions and more than 20 million people are directly or indirectly affected by the construction of dams.

There are four major projects of India which have generated major controversies:

1. Tehri Dam Project in Uttar Pradesh.
2. Sardar Sarovar Project in Gujarat.
3. Narmada Sagar Project in Madhya Pradesh.
4. Bodhghat Project in Madhya Pradesh.

Case Studies–Dams

Narmada Sagar

Narmada is the largest west flowing river arising from Amarkantaka Hills in Shahdol district of Madhya Pradesh. Construction of Narmada Sagar dam displaced over 1 million people. Land submerged about 1000 villages and over 50,000 hectares of agricultural land and forest also.

Tehri dam is constructed on river Bhagirathi. It had sunk about 26 villages and affected about 10,000 residents of Tehri town. Rehabilitation became a burning issue.

Hirakund dam displaced more than 20,000 people residing in about 250 villages. Bhakra Nangal dam was constructed in the 1950s. Till now only half of the people are rehabilitated.

Sardar Sarovar Project submerged about 20 villages.

Mining

Mining also causes displacement of people. Jharia coal fields caused underground fires in Jharkhand and people had to vacate the town. Mining in that area has caused deforestation, destruction of wildlife and other natural resources. Coal mine drainage may have up to 2,05,000 mg/l acidity with related amount of iron and dissolved solids. Air pollution levels has increased due to the production of dust particles during the process of drilling, blasting, transportation and so on. About 0.3 million population of Jharia has been displaced from that area. But rehabilitation is still a problem. About 115 crores have been spent to control the fires.

National Parks

These are the areas which are maintained by government and reserved for betterment of wildlife, cultivation, grazing, forestry and habitat manipulation. There are 85 national parks occupying nearly one per cent of geographical area. Parks are divided in three parts.

I. Core Zone: No human activity is allowed.
II. Buffer Zone: Limited activity allowed.
III. Manipulation Zone: Several activities are allowed.

In Core area tribal people started some illegal activities like smuggling, poaching, destruction of habitat of wildlife. In response to this, Kerala government has displace to the tribal community from the sancturies. About 53,472 tribal people were displaced and only 843 families have been rehabilitated till now.

Rehabilitation Policy

Following objectives have been suggested by some organizations involved in rehabilitation program:

- Land for land is better policy than cash settlements.
- New settlements should be within their own environment.
- Removal of poverty by providing land.
- Landless people, should be given assurance of employment.
- During Jawaharlal Nehru's era, five principles of tribal development was accepted as 'tribal panchsheel'.

ENVIRONMENTAL ETHICS

Ethics refers to what is right and what is wrong on a universal basis. India has a rich tradition of environmental ethics since the vedic era 5000 years ago. India has a spirit of environmental conservation ethics, inherent in our history, religion and culture.

The universe has been created and nurtured by God. Man can enjoy the beauty of nature by giving up greed. The concept of living harmoniously with environment was also practised by rishis in tapovan ashramas.

Vriksharopana concept given by poet Tagore in 1927 at Shantiniketan in 1902. But now people have changed their views. Nowadays there are two prevalent views in relation to the environment ethics. First view emphasises upon the supermacy over nature through technology, development and economic growth. Second view emphasizes upon the sustainable development.

These two views are discussed below:

First view This view is according to industrialized countries.

- Humans are the source of all values.
- Unlimited matter and energy belongs to man.
- Healthy environment belongs to healthy economy.
- The success of mankind depends upon the efficient use of resources from nature.

Second View This view is based upon earth wisdom.

- Man is not the source of all values.
- Nature does not exist primarily for human use but for all living species.
- Matter and energy resources are finite and must not be wasted.
- Production and consumption of material goods need not increase endlessly.
- A healthy economy depends upon healthy environment.
- Success of mankind depends upon cooperation with nature in the use of resources for their benefit.

CLIMATE CHANGE

Climate refers to seasonal variations, weather conditions, and extremes of weather in a region. Climate is the average weather of an area. The inter-governmental panel on climate change in 1990 and 1992 periodically gives an idea about the assessment of atmospheric abundance of green house gases and its impact on climate and recent changes in global temperature.

Climatic changes are mainly due to man made activities which disrupt the ecological balance of the environment. The concentration of green house gases the atmosphere has increased resulting in increase the temperature. Due to increase in temperature, hydrological cycle is disturbed. This results in flooding and drought in different areas of the world. This also reduces crop productivity, causing rise in sea level, reduce live stock, lead to famine and death of human being and other living organisms. Due to global warming there can be shifting of different climatic regions towards poles.

Change in temperature is not uniform all over the world. It varies from place to place. Higher latitude areas are more warmer during autumn and winter than the tropics. Poles may experience two to three times higher temperature than the average, while tropics may be warmer up to 50 to 100% than the average. The increased warming at poles reduces the thermal gradient between the equator and high latitude regions. This disturbs the global pattern of winds and currents and rainfall distribution. Ocean current shifting results in the change in climate of Iceland and Britian and make the region cool when rest of the world warms. Disturbed rainfall results in non-uniform falling of rain. Rainfall may increase the temperature which results in high evaporation rates leading to water deficit in crop fields.

Factors responsible for global air circulation:

1. Climatic changes occur due to air circulation over Earth's surface.
2. Heating of earth surface is not uniform. Air is heated more at the equator than at the poles. These differences explain why the tropical regions near the equator are hot while polar regions are cold.
3. Due to tilting of Earth's axis towards or away from the Sun, opposite seasons are occur in the Northern and Southern Hemispheres as shown in figure.

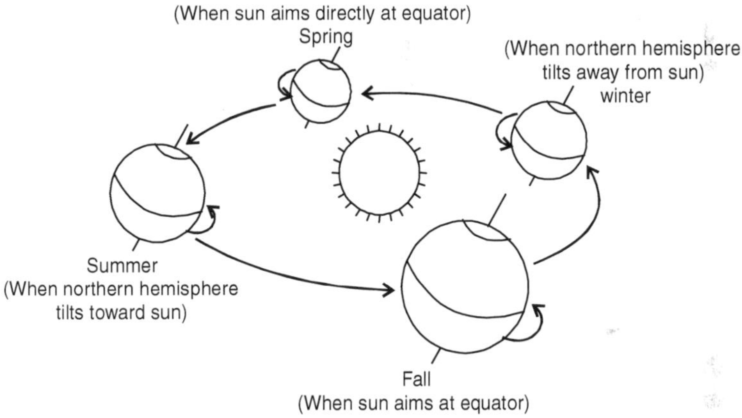

Fig.6.1 Seasons changes by the variations in the sunlight

Global warming A warming effect produced by the presence of carbon dioxide, chloroflorocarbons, methane, nitrous oxide and water vapour in the lower atmosphere is called global warming.

When the gases warm the air in the lower level of atmosphere on global scale, the effect is called green house effect. This term was coined by a mathematician Jean Fourier. The average global temperature is 15°C and in the absence of green house effect, the temperature would have been –18°C. Thus, green house effect increases the temperature to 33°C. Heat trapped by green house gases keeps the planet warm and allows other species to exist. Warming or cooling of atmosphere by more than 2°C over past few decades has proved harmful for the life on Earth.

Greenhouse gases

The gases which can absorb infrared light are green house gases. These gases increase the temperature of air on the earth.

Carbon dioxide: Carbon dioxide contributes about 55 per cent of the global warming. Industrial countries account for about 76 per cent of annual emissions.

Chlorofluorocarbons: It contributes about 24 per cent of the global warming. Main sources of chlorofluorocarbons are air conditioners, refrigerators, aerosols and evaporation of industrial solvents. Atmospheric concentration of CFC is about 0.00225 ppm and it is increasing at a rate of 0.5% annually.

Methane: Methane accounts for 18 per cent of the green house gases. Its concentration in the atmosphere is about 1.675 ppm, and increasing at the rate of 1 per cent annually. It is produced from landfills, paddy fields, digestive tracts of sheep, cattle and termites.

Particulate matter: Different size of particles are found in atmosphere. It is produced from smoke, combustion of fossil fuels, sea salts, sprays, dust from volcanic explosions etc.

Impact of Greenhouse Effect—There are number of effects on the environment due to green house gases.

(i) **Sea level rise:** Increase in global temperature leads to expansion in seawater. Due to heating, polar ice caps melt resulting in the rise in sea level. Rise in sea level may increase about 30 to 150 cm by 2050.

(ii) **Increase in global temperature:** It will increase in between 1.5 to 5.5°C by 2050.

(iii) **Effects on human health:** Global warming will result in change in temperature and rainfall which will affect the distribution of vector of diseases like mosquitoes, and result in respiratory and skin diseases.

(iv) **Effects on agriculture:** Increase in temperature leads to increase in pest growth. Evapo-transpiration will increase which may affect wheat and maize crop production.

Measures to check global warming

- Planting more trees.
- Trapping and use of methane as a fuel.

- Shifting to removable energy resources.
- Use energy more efficiently.
- Cut current use of CFCs and fossil fuels.
- Shift from coal to natural gas.
- Reduce vehicular emissions.
- Reduce the rate of population growth.
- Remove carbon dioxide from smokestacks.
- Remove atmospheric carbon dioxide by photosynthetic algae.
- Increase of nuclear power plants for electricity production.

Acid Rain Oxides of nitrogen and sulphur are the major pollutants in the air. These oxides are produced from automobile exhausts, domestic fires, power plants and smelters etc. The oxides of nitrogen and sulphur are converted into nitric acid and sulphuric acid, respectively. These combine with hydrogen chloride produced from hydrogen chloride emissions. Acid rain damages buildings which are made up of marble, limestone and mortar.

$$CaCO_3 + H_2So_4 \rightarrow CaSO_4 + CO_2 + H_2O$$

In this calcium carbonate reacts with acid to from soluble sulphates, which set leached by rain water. These acids of nitrogen and sulphur combine with water in the atmosphere and fall on the earth in the form of acid along with rain. This is known as acid rain.

Causes of acid rain

- Industralization.
- Anthropogenic sources.
- Power plants.
- Electric plants.
- Metal smelters.
- Automobiles exhaust.

Effects of acid rain

- Increased soil acidity.
- Affects flora and fauna.
- Causes acidification in lakes and streams.
- Affects crop productivity.
- Affects aquatic flora and fauna.
- Affects human health.
- Corrodes monuments, buildings, statues, buildings, railings etc.

- Levels of heavy metals increase in water bodies.
- Fish population is affected in water bodies.
- Ecological imbalances occur due to killing of bacteria and blue green algae.
- Forests are affected due to weakening of trees.

Control of acid rain

- Control of liming of lakes and soils.
- Emissions of sulphur and nitrogen oxides should be reduced by using proper pollution control equipments.
- Stricity enforcing the environmental acts.

Ozone Layer Depletion In the stratosphere, the second region of atmosphere, ozone layer is present for the last 450 million years. This ozone layer acts as a natural sunscreen for protecting the living organisms from the harmful effects of ultraviolet radiation. Ozone strongly absorbs ultraviolet radiation ranging from 295-320nm. But recently due human activities dangerous chemicals are issued in the atmosphere, which reactruith ozone and reduces its concentration. This leads to thinning of ozone layer and a ozone hole develops.

Thinning of ozone layer: Discoveries related to thinning of ozone layer were made by Shewood Rowland, Paul Crutezen and Mario Molina. It was first discovered in 1985 over Antartica. The existence of ozone hole was confirmed above Arctic in 1990. The thickness of ozone hole is measured by Dobson spectrometer and expressed in Dobson unit.

Pollutants responsible for thinning ozone layer

- ChloroFluoro Carbons
- Methane
- Nitrous oxides

Effects of ozone depletion

- Increase in incidents of cataract.
- Melanin producing cells get destroyed.
- Death and decay of phytoplanktons, zooplananktons, fishes etc.
- Reduces yield of crops.
- Degradation of paints, plastics and polymer materials result in economic loss.
- Inhibit photosynthesis in plants.

- Damages nucleic acids in living organisms.
- Decline in functioning of immune system.
- Causes skin cancer.

Control of ozone depletion

- To limit the production and use of ozone depleting substances.
- Use of alternatives of CFCs.
- Phasing out ozone depleting substances.
- Establishing the principles for reducing green house gas emissions.

Nuclear Accidents and Holocaust Nuclear accidents have disastrous impact on human society. Nuclear war can destroy the entire world and biological community due to nuclear holocaust within no time. It can occur at any stage of the nuclear fuel cycle as well as due to leakage of radioactive materials. The nuclear bombs are about 10 megatons and 500 kilotons. Nuclear bombardment will cause combustion of wood, petroleum, plastics, forest etc. A black soot is produced by these and released into the stratosphere. This soot absorbs solar radiation and would not allow the radiation to reach Earth resulting in cooling effect. Due to cooling effect, water evaporation will get reduce and this phenomenon is known as nuclear winter. It is just opposite to global warming because nuclear winter lowers down the global temperature in summers as well. This reduces the yield of crop and causes famines and human suffering.

Case Study

The worst nuclear disaster occurred in 7 October 1957 in Soviet Union in Southern Ural mountains around the city of Kyshtym. In this, massive amount of radiation was released. Due to which area became deserted, river was diverted, 30 towns and villages disappeared. In 1945 Nuclear bombs exploded in Hiroshima and Nagaski of Japan. This explosion killed about 100,000 people and destroyed the city. The explosion emitted neutrons and gamma radiation. In 1986 Chernobyl nuclear accident resulted in spread of contamination of radioactive substances.

Wasteland and Reclamation Wasteland is defined as land that is at present lying unused or land, which is incapable of producing material or service of some value. In India about 175 million hectare of degraded land is present. Wasteland amounts to be about 5.35 per cent of total geographical area of Haryana (INDIA). Thus, indicating an acute problem for India, which has to support about $1/6^{th}$ of the world population. Wasteland includes degraded forests, Overgrazed pastures, drought struck pastures, eroded valleys, hilly slopes, and water logged marshy areas etc.

Causes of wastelands

Following are the causes of wastelands:

(i) **Natural causes:** Due to action of wind, water and weather degradation of land occurs. This can be checked effectively by action oriented strategies.

(ii) **Economic causes:** Lack of funds also cause wastelands to develop.

Wasteland reclamation

There are about 87 million hectares of agricultural land which seems to be degraded by severe soil erosion. Reclamation of wasteland involves high expenditure, expertise and manpower.

The various measures for wasteland reclamation are as follows:

1. **Afforestation:** Afforestation means growing forests over wastelands. Trees used as windbreaks are planted in deserts. Examples are *Calotropis, Zizypus, Jujube, Acacia, Dalbergia sisso,* and *Tamarindus* serve as useful windbreaks.
2. **Changing agricultural practices:** Mixed cropping and crop rotation have changed agricultural practices which improves fertility of wasteland.
3. **Leaching:** Saline lands can be recovered by providing water, so that leaching of salts take place.
4. **Mulching:** Mulch is the protective layer formed by the stubble, which is the basal part of herbaceous plants or remains attached to the soil after harvesting. Mulches not only act as wind barrier but also reduces evaporation and increases soil moisture.
5. **Reclamation of saline soil:** Strategy for reclamation of saline soil has been adopted by the use of gypsum, garbage, green measure, paddy, straw etc. for 2-3 years. Use of gypsum and pyrite is more superior in reclaiming the saline soils.

6. **Managing topography:** When water runs downhill, it erodes soil, and carries off the fields. Water runoff can be reduced by strip farming, tied ridges and terracing.
7. **Providing ground cover:** Ground cover reduces evaporation and soil temperature in hot climates and protects ground organisms that helps in aeration and rebuilding of soil.
8. **Gully control:** Gully control can be done by constructing dams, drains etc.
9. **Stream bank protection:** Stream bank protection can be done by growing vegetation or by providing concrete or stone pitching etc.
10. **Development of industries or residential areas:** Wasteland areas should be preferred for the setting up of industries. The wastelands in India approximately 55 per cent as compared to cultivable area. Therefore there is a need of proper wasteland reclamation.
11. **Drainage:** This method is applied for water logged soil reclamation, where excess water is removed by artificial drainage for controlling soil salinity.
    (i) **Surface drainage:** This is used in areas where water stands in 30-45 cm deep ditches in the fields lying parallel to each other at 20 to 50 m, distance. This way 5 cm of water can be removed within 24 hrs rains by providing ditches to runoff the excess water.
    (ii) **Sub-surface drainage:** Horizontal sub-surface drainage is provided in the form of perforated PVC pipes or open jointed pipes below the land surface. The world bank has funded sub-surface drainage system at Sampla, Rohtak (Haryana).

Consumerism and Waste Products

Consumerism means consumption of resources by the people. Early man used to consume less resources. But due to industrialization, consumerism has shown exponential rise. Hence, the modern society have multiplied our needs and so consumerism of resources has also increased. For thousands of years our population was even less than 1 million. But now we have crossed 5 billion and by 2045 it would cross 11 billion. Now let us see how the changing population influences consumerism, and waste

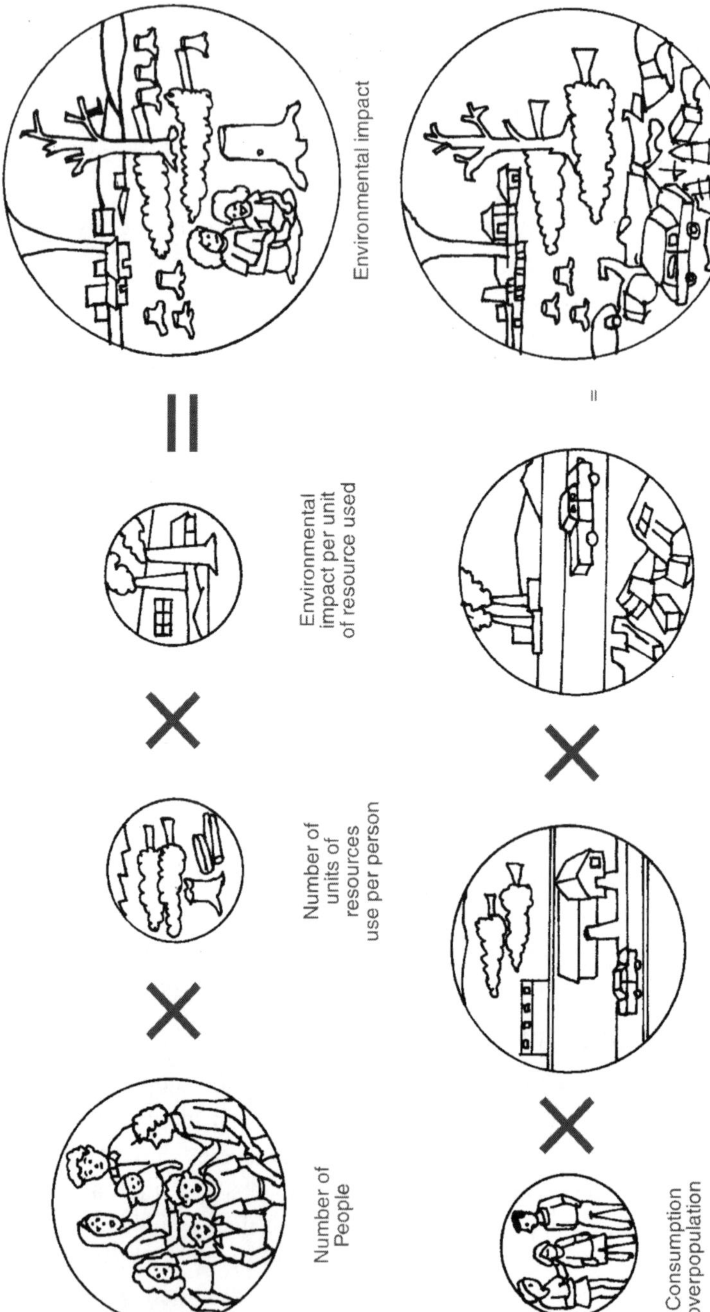

Fig. 6.2 Two types of overpopulation depends upon the relative importance of the factors in the model shown in figure.

products. Here we will discuss two types of conditions of population and consumerism.

1. **People over-population** This type of condition occurs when there are more people than the supply of food, water and other required resources in the area. This leads to poverty, under nourished people and premature deaths. This type of condition occurs in less developed countries (LDCs). So there is less per capita consumption although overall consumption is high.
2. **Consumption over–population** This type of condition occurs where population size is smaller than the resources and due to luxurious life style, per capita consumption of resources is very high. More the consumption of resources, more is the waste generation and greater is the degradation of environment.

This can be shown by using the Model of Paul Ehrlich and Johu Hodlreu (1972).

Number of people + Per capita use of resource + Waste generated per unit of resource used = Overall environmental impact.

From the above model following conclusions can be drawn

- In a less developed country the population is high but per capita use of resources and waste generated is less.
- In not so well developed countries the population is low, but per capita use of resources and waste generation are higher.

So overall environmental impact consumerism may be greater in not so well developed countries.

Table 6.1 Comparison of (USA and India) consumerism and waste generation

Parameter	Per cent global values	
	USA	India
Production of goods	21%	1%
Population	4.7%	16%
Pollutants/wastes	25%	3%
CFC's production	22%	0.7%
Energy use	25%	3%
Global production	18%	4%

Three factor model given by Paul Ehrlich is important for understanding the factors leading to unsustainability. But here the interconnected environmental resource problems and social problems are more complex and involve a number of poorly understood mix of interacting factors are shown in figure.

STEPS TO SUSTAIN ENVIRONMENT

Environment Protection Act 1986

This act provides protection and improvement of environment. This act was enacted by the parliament in the 37th year of republic India in 1986. This act extends to the whole of India. Few terms related to environment are described in the act are as follows:

- **Environment:** It includes water, air, land and inter- relationship, which exists among and between water, air and land and human beings and other living creatures.
- **Environment pollutant:** It means any solid, liquid or gaseous substance present in such a concentration as may be or tend to be injurious to health.
- **Environment pollution:** It means presence of any environmental pollutant in the environment.
- **Hazardous substances:** It means any substance or preparation, which by reason of its chemical or physico-chemical or handling is liable to cause harm.
- **Occupier:** A person who has control over the affairs of the factory or the premises.
- **Prescribed:** Prescribed by rules made under this act.

This act has given powers to central government to take measures to protect and improve the environment in the following ways:

1. Establishe laboratories
2. Execute nation wide program for prevention and control of environmental degradation.
3. Examine the manufacture apprises and the premises.
4. Lay down standards for quality. Stress on emission standards.
5. Lay down procedures and safeguards.
6. Sponsor researches for environmental problems.

Central government has power to make rules to regulate the environmental pollution in any of the following area.

1. To set the standard of quality of air, water and soil.
2. To set the minimum allowable limit.
3. The procedures and safeguards for handling hazardous substances.
4. The procedures and safeguards for the prevention of accidents.
5. The prohibition and restriction on the location of industries and to carry on the process

For the prevention, control and abatement of environmental pollution, standards have been specified under schedule-I-IV for emission of gaseous pollutants and discharge of effluents from industries on operations.

Under the Environmental (protection) rules, 1986 the State Pollution Control Boards have to follow the guidelines provided under schedule VI, some of them are discussed below

1. To advise the industries for wastewater treatment and discharge of gases within the standard limit.
2. To advise recycle and reuse of wastes.
3. The Central and State Boards should give emphasis on the implementation of clean technologies by the industries in order to increase fuel efficiency and reduce the generation of environmental pollutants.

There was an amendment in 1994 for environmental impact assessment of various development projects. These projects require clearance from central government before their commencement. In 1989 there was also an amendment that included about hazardous waste management and handling. Under these rules, it is the responsibility of the industrialist to take all measures for disposal of wastes without any adverse effects.

The Environment Protection Act 1986 has also made an amendment regarding environmental audit as a means to check whether an industrialist is complying with the environmental laws and regulations or not.

Water (Prevention and Control of Pollution) Act, 1974 This act is for the prevention and control of water pollution and maintaining or restoring the utility of water. Following are the features and provisions of the Act:

1. It provides for the establishment of central and state boards for pollution control.
2. It provides for maintenance and restoration of quality of surface and ground water.
3. It confers them with powers and functions to control pollution.
4. The state pollution control and central boards are given powers to advise, coordinate and provide prevention and control of pollution of water.
5. This act has provisions for funds, accounts, budgets, and audit of the central and state pollution control board.
6. This act makes provisions for different penalties for the defaulters and procedures for the same.

Powers given to central and state pollution control boards for prevention and control of water pollution are as follows:

1. State board have powers to obtain information of any area.
2. Board has powers to take samples of effluents.
3. Board has power of entry and inspection.
4. Organize Programmes (British SP) on pollution through mass media.
5. Establishes laboratory for analysis of water and effluents.
6. Board may prohibit the use of stream or wells for disposal of polluting matter.
7. Board maintains proper accounts and relevant record for annual statement.
8. Central government has powers to supersede the central board and joint board.
9. State government has powers to supersede the state board.

The act has made provisions regarding the power of the boards to obtain information, enter and inspect the units and sanction or refuse consent to the industry after effluent analysis.

Air (Prevention and Control of Population Act 1981)

1. This act provides for the prevention, control and abatement of air pollution.
2. This act was called Air Prevention Control Act in 1981 and extended to whole of the India.
3. This act defined various terms like air pollutants, air pollution, approved appliances, approved fuel, board, central board,

chimney, control equipment, emission, industrial plant, member, prescribed, occupier, and state board.

4. This act is controlled by state board and central control pollution board and functions similarly as in water prevention and control of pollution.
5. Central board and state board controls the quality of air.
6. It plans for a nation wide programme for the control or abatement of air pollution.
7. It organizes training for improvement of air quality.
8. Lays down standards for quality.
9. Establishes laboratories.
10. Similar to Water Act. The Act has provisions for powers, functions of pollution control board, funds, accounts, audits, appeals, penalties and procedures.
11. Board can declare an area as air pollution control area.
12. Operation of industrial plant is with the consent of the board.
13. Emission of air pollutant is within the permissible limit in air pollution control area and is controlled by the board.

Wildlife Protection Act 1972 Wildlife legislation was started in 1972 and the Indian Board of Wildlife was established in 1952. Following are the provisions for the Wildlife Protection Act 1972.

1. This act defines the terms related to wildlife.
2. This act involves the appointment of wildlife warden and advisory board and their powers and duties.
3. This act involves prohibition of hunting of endangered species.
4. This act provides facilities for the establishment of wildlife sanctuaries and national parks.
5. This act provides ban on the trade of wild life.
6. It provides breeding programs for endangered species.
7. It provides provision for trade of some wild life species.
8. This act provides for the constitution of central zoo authority.

Forest Conservation Act 1980 This act provides for the conservation of forests and it was enacted by the parliament in 1980. This law extends to whole of India except Jammu and Kashmir. Following are the features and provisions:

1. This act provides the use of forestland for non-forest purpose.
2. This act provides appointment of advisory committee by central board.
3. This act provides provision for penalty of contravention.
4. This act provides punishment when any authority and government department commit offences.

Issues Involved in Enforcement of Environmental Legislation

Environmental legislation is for safeguarding our environmental quality. But even then we are not able to achieve our targets. This may be due to drawback in environmental legislation and problems in their effective implementation.

Following are some of the issues related to the acts:

1. **Forest Conservation Act, 1980**

   In this act, there is just transfer of powers from state to center. Tribal or local people have kept out from the decision making process regarding the nature of the use of forest area. Their role is neither acknowledged nor honored. Now efforts are on to patch up the loopholes in laws by introducing the Human Rights Protection.
2. **Wildlife Protection Act, 1972**

   Offender of the act is not subject to harsh punishment penalties.
   - Opening avenues for illegal trading in Jammu and Kashmir also serves the purposes of trading.
   - Ownership certificate for animal articles also enhances illegal trading.
3. **Pollution Related Acts**
   - Penalties on treatment plant and pollution control equipment is less.
   - Illegal mining is not under control in Rajasthan due to centralization of power at State level.
   - State boards clarify adequate funds and expertise to pursue their objectives.
   - The act should make some subsidies for installing treatment plants.

- Position of chairman of the boards is occupied by politicians rather than an environmental expert.
- Public involved in decision making with ministry of Environment and Forests (1992) is yet to be implemented.

Hence, environmental legislation should be aimd at following

- Community state partnership.
- Administrative transparency.
- Strong penalties to offender.
- Decentralization of power.
- Environmental law education.

Public Awareness

The Bhopal gas disaster in 1984 and Earthquake in Indian Ocean on 26 December 2004 occured due to negligence and lack of awareness on environment. People must be made aware about both formal and non-formal education. Formal education is for the students in the educational system. Non-formal education is for the non-educational institutions. The University Grant Commission recently introduced syllabus on Environment for the undergraduates in all streams.

Public awareness can be created through following methods:

1. **Among the people through mass media:** Public awareness can be created by radio, newspaper, magazines, telling environ-mental conservation stores, television programme audio and video CDs and cassettes etc.
2. **Among students through education:** Awareness can be created among students by teaching environment as a compulsory subject from primary classes to graduate classes.
3. **Among the farmers and local people:** The seeds of environmental awareness can be effectively propagated by organising training camps.
4. **Among the politicians and beuracrates:** Politicians and beuracrates can play an important role in bringing manures in the society. These people should attend orientation courses or workshops related to environmental awareness.
5. **National Policy on environment:** Ministry of environment and forests publishes a magazine related to latest deve-lopments in

environment. The worldwide Fund for Nature WWF, Center for Science and Environment, CSE and Bombay Natural History Society, BHNS, are all government organizations and play a significant role in environmental awareness through research and other activities.

6. **Non-government organizations:** Non-government organisations play a significant role in creating a link between government and local people for addressing environmental issues and generating an awareness. There are about 200 non-government organisations out of which 130 are engaged in awareness programme, 50 in pollution control, 50 in nature conservation, 45 in afforestation and social forestry, 15 in rural development and 10 in ecodevelopment. All these are academically oriented activities. Kerala Sastra Sahitya Parisad, KSSP is a non-government organisation and has piloted the silent valley movement.
7. **Organization of environment movement:** In critical situations when environment degradation takes place, environmental movements are organised such as Chipko Movement and silent valley movements and so on.

If you want to educate people and make them aware of environment conservation, then it is rightly said "**If you want to act green, first think green**".

REVIEW QUESTIONS

Short Type Questions

1. Define sustainable development.
2. What do you mean by water conservation?
3. What is watershed management?
4. What do you mean by the term rainwater harvesting?
5. What are the factors responsible for rehabilitation?
6. What do you mean by environmental ethics.
7. What are the effects of global warming?
8. What causes global warming?
9. What are greenhouse gases?
10. What is green house effect?
11. How can we prevent global warming?
12. What is acid rain?

13. What causes acid rain?
14. What are effects of acid rain?
15. How can we prevent acid rain?
16. What is ozone layer depletion?
17. How is the ozone layer is the thinning?
18. How can we prevent ozone layer depletion?
19. What are the effects of ozone depletion?
20. What is meant by the term wasteland?
21. What causes wastelands?
22. Name some government organization related to environment.
23. What are environment pollutants?
24. Name some non-government organizations related to environment.

Descriptive Questions

25. What are the main features of sustainable development?
26. How sustainable development can be achieved?
27. Mention few urban problems related to energy.
28. Discuss in detail the methods of water conservation.
29. What are the methods for watershed management?
30. Name some methods for rain water harvesting.
31. What are the aims of rain water harvesting?
32. Discuss some problems related to resettlement and rehabilitation with suitable examples.
33. Discuss in detail the environmental ethics.
34. How climate change affects the environment?
35. How can we control global warming?
36. Explain in detail the nuclear accidents and their effects in environment.
37. How wasteland can be reclaimed?
38. Discuss consumerism and waste products in relation to developed and developing countries.
39. What is environment protection act 1986 and discuss it in detail?
40. Explain water (prevention and control of population) act 1974.
41. Discuss Air (prevention and control of pollution) act 1981.
42. What is wild life protection act 1972?
43. Discuss forest conservation act 1980.
44. What are the issues involved in enforcement of environmental legislation?
45. How can public awareness be created among people?

INTRODUCTION

Population of our ancestors was confined to Africa and numbered only in lakhs. The invention of agriculture, hunting, gathering of food and identification of economics helped in expansion of growth of population.

The industrial revolution in Europe and North America created a new society in which primary activities like agriculture, forestry and fisheries began to decline and the secondary (manufacturing) and tertiary (services) activities began to take an increasingly important role. This led to diversity in national economy. Due to industrial revolution, urban population increased rapidly.

Table 7.1 World: Size and Growth of Population

(In million)

World/ continent	1950	1998	2025*	2050*
World	2523.87	5929.83	8039.13	9366.72
Asia	1402.02	3588.87	4784.83	5442.56
Europe	547.31	729.40	701.07	637.58
Africa	223.97	778.48	1453.89	2046.40
Latin America	149.29	462.59	641.40	754.20
Anglo-America	171.61	304.07	369.01	384.05
Oceania	126.12	294.60	40.68	45.68

* Projected Figure

Source: World resources, 1998-99 World Resources Institute, UNO, New York.

Thus world population has increased in the last few centuries. But the rate of increase quickened only after 1900. The growth of population may be appreciated from the fact that in 1950 the total population of the world was more than 2,523 million, which rose to more than double in 48 years being 6 billion in 1999. The estimated population of the world at the present rate of growth would be about 8,039 million in 2025 and about 9,366 million 2050.

From the Table 7.1, it is observed that population growth rate would be highest in Asia, followed by Africa and Latin America. There would

7

Population Growth and Variations among Nations

CHAPTER OBJECTIVES

- Introduction
- Environment and Human Health
- Value Education
- AIDS
- Human Rights
- Role of Information Technology in Environment and Human Health

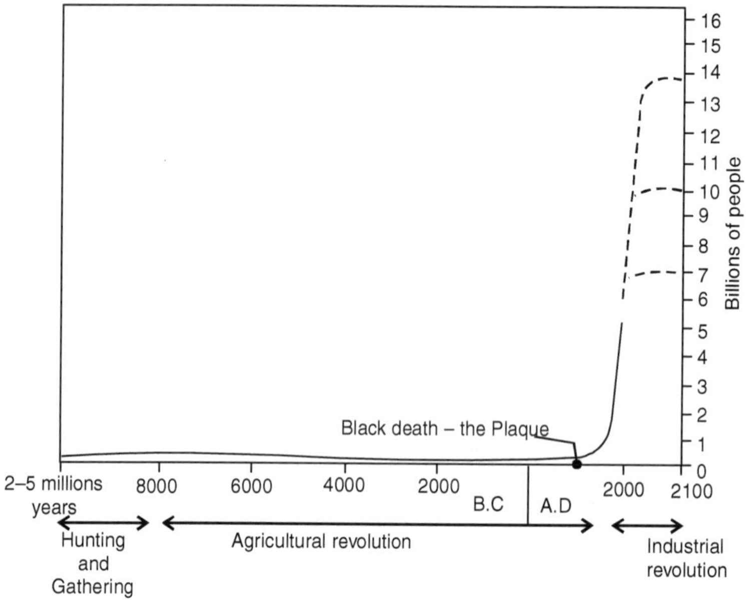

Fig. 7.1 J shaped curve showing post exponential world population growth (Data from World Bank and United Nations)

be a marginal increase in the population of Anglo-America (USA, Canada) and Oceania. While the population of Europe will decline from 729.40 million in 1998 to about 701.07 million in 2025 and 637.59 million in 2050. The negative growth of population in Europe is of great concern. Most of the European countries have adopted the policy to give incentives to the people to increase population.

United Nations divided the world's countries in more developed countries (MDC's) and less developed countries (LDC's). MDCs have high gross national product GNP per capita and are highly industrialised. Such as Canada, Japan, Australia, New Zealand, United States and Western European Countries. All the above MDCs have about 22 per cent of world population and utilize about 85% of world wealth, 73% of energy, 88% of natural resource and produce about maximum pollution and waste.

While the less developed countries (LDCs) have low industrialisation and GNPs per capita. These LDCs countries have about 78% of world population and consume about 15% world's wealth, 27% of energy and 12% of natural resources LDCs countries are Indonesia, Malaysia, Brazil, Mexico, Thailand and Chile. So world population increase takes place in LDCs that is about one million people are added in every four days.

Types of Population Growth

Population growth is increasing in its size due to increase in number of organisms. It can be determined the number of individuals added to the population and the number of individuals lost from the population.

There are three types of population growth:

- **Positive growth:** If more individuals are added to the population. The population shows positive growth.
- **Negative growth:** If more individuals are lost from the population. The population shows negative growth.
- **Zero growth:** When addition or loss of individuals balanced. Then the population shows stationary or zero growth.

Population growth is regulated by biotic potential and environmental resistance factors. Population growth is also represented by population growth curves. It is the mathematical expression of the growth of a population from its beginning till it stabilizes.

There are two types of growth curves or sigmoid.

- **S Shaped** - This type of growth curve is shown by animals as well as by yeast cells grown under laboratory conditions.
- **J Shaped growth curve** - This type of growth curve is shown by animals or those animals which feed on seasonal plants.
- **J Shaped curve of Man** - The increasing size of human population is an example of exponential growth. Initially growth curve creates a horizontal curve but later on growth speeds up. As the human population grows, the number of people on earth soars and population growth curve shows a bend and heads and then straight up and shows J shaped curve as shown in figure 7.1.

Every year, the world population grows by almost 90 million due to increase of 2.8 people each second. It may be observed from the table that China's population ranks first followed by India, U.S.A., Indonesia, Brazil, Pakistan, Russia, Bangaladesh and Japan.

Table 7.2 Size, growth and doubling period of population (2000)

Country	Population (million)	Years for population to double
China	1277.0	61
India	1027.0	35
USA	281.4	100
Indonesia	212.1	43
Brazil	170.1	40
Pakistan	156.9	27
Russia Federation	146.9	ND
Bangladesh	129.2	29
Japan	126.3	270
Nigeria	111.5	26
Mexico	93.0	31
Germany	81.2	ND
Iran	62.2	25
Egypt	60.1	31
U.K.	58.6	280
France	58.0	182
World	6055.0	43

Sources: Census of India 2001.

Table 7.3 Size growth and percentage of population

Country	Population in Million			
	2000	1%	2025	%
China	1277	21.50	1480	21.14
India	1020	16.87	1330	19.38
U.S.A.	281.4	5.00	332.4	4.74
Indonesia	212.1	3.50	275	4.07
Brazil	170.0	2.80	216	2.80
Pakistan	156.5	2.60	268.9	3.93
Russia	146.0	2.43	131	1.99
Bangladesh	129.2	2.15	180	2.96
Japan	126.9	2.11	121	1.75
World	6000.0	-	8039	2.96

Source: Census of India 2001

Factors responsible for the variations among the nations:

1. Availability of arable land and water
2. Age of civilization
3. Accessibility of places
4. Restrictions of national boundaries

Table 7.4 Growth rate of population of developed and developing countries

Country	Life expectancy at birth	Annual growth rate of 2001 (%)
Bangladesh	58	2.0
Nepal	57	2.1
India	69	1.6
Pakistan	70	2.4
Indonesia	67	1.1
Nigeria	56	2.2
China	70	0.7
South Korea	72	1.0
U.K.	77.2	0.0
Japan	80	0.0
France	79	0.3
Germany	77.2	-0.01
U.S.A.	77	0.7
Russia	67	-0.2

Source: Census of India 2001

Population Explosion—Family Welfare

Human population is increasing at the rate of 2% a year and becomes double every 35 years. In 1965 the human population was about 3.5 billion and reached 7 billion in 2000 and 8.4 billion by they year 2005 and 11.2 billion by the year 2010. This high rate of human growth is termed as population explosion.

Population explosion is one of the strongest problems in India. India is the second most populated country of the world. In 1991 the population of our country was 84.63 crores and in 1999 it was about 98.13 crores. China has the largest population in the world but the net addition of population is lower than India. China's population is crossing by 1.24 crores/year. While India's population is increasing by 1.88 crores/year.

The increase in population growth is due to various reasons. The main causes are changes in the mortality and fertility conditions of the population. The difference between the two is the main cause of rapid population growth in our country.

Factors responsible for increase in population:

Mortality: The main cause of increase in population is reduction in mortality rate. This is due to improvement in medical technology, and increase in health status of the population. The crude death rate was 27.4 in 1941-51, which decreased to 8.9 in 1997. It means that for every 1000 population, 27 persons were dying in a year in 1945, while only 9 persons die per 1000 population at present.

Fertility: Level of fertility is mainly responsible for the population growth in India. Level of fertility may be measured by Crude Birth Rate CBR. CBR indicates the number of live births per 1000 remained at a high level with a very slow pace of decline even in the post independence period. It is only since the eighties that the pace of decline has somewhat accelerated as compared to the previous periods.

The difference of Crude Birth Rate (CBR) and Crude Death Rate (CDR) indicates the natural growth rate in a population. With a large difference between CBR and CDR, a large number of persons are added to the population and results in population explosion. Crude Death Rate decreases sharply during the last years which means the number of aged persons are more while the decrease in crude birth rate was not so sharply by which number of healthy infants increasing which resulted in a large gap between the CBR and CDR. As a consequence, each year increasing number of persons were added to the population in terms of birth and less number of populations removed from the population in terms of death. This greater addition of persons resulted in population explosion.

Usually transition occurs in mortality and fertility conditions in the population of all countries. This is known as "Demographic Transition".

However, the time taken for these changes in birth and death rate is very important in terms of population explosion. Longer time involved in completing the demographic transition produces great problems of population explosion due to longer period of high fertility regime.

The level of fertility of a population is directly connected with the extent to which the eligible couples adopt any method of family planning. The percentage of couples using family planning method is only 36 per cent according to National Family Health Survey (NFHS) in India (1992-93) and it went up to 60% by the year 2001. In India, terminal methods are emphasized for family planning. The temporary methods acceptance is very little among couples.

Unmet need for family planning is that when a married women does not want any more children or wants to wait for two to three years before having another child, but is not using contraception. In India there are 20% who are suffering from unmet need for family planning. Unmet need for family planning has been measured and results in estimation of demand for family planning in the country, which could not be provided by the country's family planning programme.

India framed "National Population Policy" in April 1976. It emphasized on the importance of the small family norm, offer better services and include newer contraceptive methods. India was the first country to launch an official family planning program *i.e.* "Cafeteria Approach" in 1952 which was later shifted to 'extension' and 'camp' approaches. Now emphasis has been on reproductive and child health RCH approach.

Family planning is defined by WHO as "A way of thinking and living that is adopted voluntarily, upon the basis of knowledge and attitudes and responsible decisions, by individuals and couples in order to promote the health and welfare of the family groups and thus contribute effectively to social development of a country".

Family planning helps everyone as shown below:

Women

1. Family planning prevents high-risk pregnancies.
2. Family planning prevents unwanted pregnancies.
3. Family planing prevents unsafe abortions.
4. Family planning helps women's body to replenish the lost stores of nutrients.

Men

1. Family planning helps them to care for their families.
2. Family planning helps men to provide a better life for their families.

Families

1. Family planning makes people economically viable.
2. Family planning is able to provide children with food, clothing, housing and education.
3. Family planning provides family well being.

Nation

1. Family planning helps in nation's development.
2. Family planning helps in faster economic, educational improvement.

Earth

1. Family planning ensures less demand on natural resources.
2. Family planning helps in better opportunities for better lives.
3. Family planning helps in less differences in population between developed and developing countries ensuring a more balanced world.

ENVIRONMENT AND HUMAN HEALTH

Health of any individual or nation is determined by the integration of two ecological universe *i.e.* internal environment of man himself and external-environment, which surrounds him. The environment has a direct impact on physical, mental, economic and social well being of those living in it. No one can afford to neglect environment. The problems of environmental health are many so, it is important to save our planet, for the present as well as for the future generation.

Environment is the air, water, minerals, organisms and all other external factors surrounding and affecting a given organism at any time. World health organization in 1948 gave the definition of human health. "Health is a state of complete physical, mental and social well being and not merely an absence of disease or infirmity". So it recognizes four dimensions of health – mental, social and physical and spiritual.

Mentally Healthy person is one who has following characteristics:

- He neither underestimates nor overestimates others.
- He has moral or spiritual values too.

- Has the ability of facing stresses and strains.
- He can think for himself as well as for others.
- He is able to live in harmony with others.

Physical health refers to normal appearance – structure, size, and functions of all the parts of the body. Social health refers to the sum total of all programmes aimed at maintaining, protecting and improving the health of the people.

Ways to Protect Our Health by Reducing Environmental Degradation

It is our moral duty to protect our environment at different levels-

At Individual Level

- Use cycle.
- Save petrol.
- Do not eat junk food.
- Grow more trees.
- Quit smoking.
- Don't blow horn continuously.
- Use tube lights.
- Use public transport.
- Be vegetarian.
- Before leaving the room, switch off all the switches.

At Family Level

- Grow more trees.
- Use eco-friendly refrigerator.
- Use unleaded petrol.
- Use public transport.
- Use Solar Equipment.
- Grow indoor plants.
- Do not use more flush water.
- Do not use polythene bags.
- Reduce use of detergents.
- Use dust bins.

At Community Level

- Grow more trees.
- Use unleaded petrol.

- Restrict use of loud speaker.
- Minimum use of crackers.
- Minimum use of pesticides and insecticides.

At last, always remember four R's

- Reuse
- Recycle
- Reduce
- Repair

The Ministry of environment and forest, Govt. of India has instituted a scheme of labeling environment friendly products. This scheme is known as eco-mark scheme. Eco-mark scheme is the scheme in which products are less harmful to environment. Eco-mark is given to eco-friendly products.

Any product which is made, packed, used or disposed off in a way that significantly reduces the harm it would otherwise cause to the environment is considered as an environment friendly product. Products, which are bio-degradable are more environment friendly than those which are not degradable.

VALUE EDUCATION

Value can be defined as a belief, which guides human behaviour and helps in making decisions about choices. Values are considered as desirable and important, by the society in which a person resides. Value shows personal attitudes, decisions, choices, judgement, behaviour, relationships, dreams and vision. Values also influence our thoughts, feelings and actions. They provoke us to do right things. Values are the guiding principles of life. They give directions and firmness to life. Values are like the rails that keep a train on the right track. They bring joy, satisfaction, peace and quality to life.

Concept of Value

Concept of value education is very extensive as it includes morals, religion etc. It consists of all the component parts like individual, society, nation universe and so on. Meaning of value is related to economic value, philosophers like Rudullah Lotse in nineteenth century attached more extensive meaning to value. In broad sense, we use the term value as literary value, democratic value, life value and education value in our day to day speaking and writing.

Components of Value Education

- Moral Education
- Environmental Education
- Population Education
- Human rights and Duties
- Health Education
- Indian Culture
- Physical Education
- Yoga Education.
- History of Indian freedom movement.

Education plays a vital role in increasing the capacity of the community for good quality of life. Education also helps in sensitizing one's awareness and for motivating one's behaviour to suit rising needs and demands for survival, growth and development. It's aim is to achieve the goals of progress and prosperity. It can only erode ignorance. Education plays a vital role in increasing the capacity of the family and community for a better quality of life.

The present education is the process, which makes a person to develop into a human being. The final result of education is the development of cultured person. The present education system does not guarantee a value based life.

The growing concern over the erosion of essential values and an increasing criticism in society has brought to focus the need for readjustment in the curriculum in order to make education a forceful tool for the cultivation of social and moral values.

Value education helps to eliminate obscurantism, religious fanaticism, violence, superstition and fatalism. Value education should lay primary emphasis on "Everyone has the right to education". To achieve this goal following principles are laid down:

- Education shall be free at elementary stages and should be made compulsory.
- Technical and professional education shall be made generally available.

- Higher education shall be equally accessible to all.
- Education shall undertake activities for the maintenance of peace.
- Parents have a right to choose the kind of education that shall be given to their children.
- Education shall be directed to the full development of human personality.

Value education should emphasize on the following points:

- To develop the ability to distinguish between the right and wrong.
- To live in harmony with nature and world.
- To develop an attitude of genuine appreciation for the value and respect for all forms of life.
- To be aware of social problems.
- To be ready to accept leadership.
- To make an individual responsible towards society, family and his surroundings.
- To contribute towards intellectual, moral, physical and personality development.

Value education constitutes a centrifugal force with full control over the objection, content, methodology and management of education. The benefits of education should reach beyond the classroom to the adult population.

Need of Value Education

If we analyze the Indian social position, it has been found that traditions of Indian culture are degrading day by day. The reason for degradation might be increasing in-discipline and destructive mentality. Among the special features of education, moral and basic education occupies important place. Indian education system was lacking social, spiritual and moral values in the curriculum of secondary education level.

The ancient education system in India was full of religious impressions. The expert persons were very conscious about religious education and they suggested moral education in place of religious education. India is a country of varieties. So Indian education commission suggested the curriculum based on social, spiritual and ethical values. Almost all the commissions and committees recommended value education. National Education Policy (1986) however stamped on value education.

Value education teaches us to preserve good things and to develop moral, social, aesthetic and spiritual sides of a person which are undermined in formal education. It also develops sense of humanism, a deep concern for the well being of others and the nation. It helps us to accept and respect the attitude of those who differ from us. It does not mean value imposition or indoctrination. Value education has the capacity to transform a diseased mind into healthy mind. This leads to fulfillment of the evolutionary role in man.

Gupta (1988) gave views about value education and explains that value education is practicable and wider than religious education, as no faith or religion is reflected through ethical, moral, social and cultural values. This means that value education is instruction in the entire realms of values – physical, emotional, intellectual, aesthetic, scientific, democratic and moral for any individual.

According to Subramanyan (1990), value education is the concept and practical implications of God's prayer, truth, dharma, love, action, religion and grace. Roy Chaudhry (1989), Reddy & Reddy (1992) and Bhat (1996) explained the concept of value education relevant to the elementary, secondary, college and university levels; the objective being the development of an individual, character building and self-respect.

AIDS

AIDS was first detected in U.S.A. It might have started in 1940's and 1950s. The meaning of AIDS is **Acquired Immune Deficiency Syndrome**. The NACO National Aids Control Organization reports that 2.9 million people are affected with AIDS and 42,527 are tested positive for the human immuno-deficiency virus (HIV) disease Nation wide. Of then 3,161 had progressed to AIDS in which virus cripples their immune systems and then leads to tuberculosis and cancer.

In 1986 United Nations program on HIV AIDS estimated that India has 3 million HIV infected people, which are the largest in the world population. South Africa has 1.8 million, Uganda 1.4 million and Nigeria has 1.2 million. In 1995 WHO reports estimated that, 30.6 million people were infected with AIDS globally.

In India, the dominating states are Maharashtra, Tamil Nadu, Manipur and Delhi. The people infected with AIDS fall with in 20-40 age group. The infection is maximum among the STD patients, haemophilic, thalassemic, homosexuals and the persons having sexual contacts with more than one partner.

Table 7.5 Sources of Aids Infection

Sources	No. of people	Percentage
Blood transfusion	216	6.8
Homosexual contact	24	0.8
Blood product infusion	25	0.8
Hetrosexual promiscous	2333	73.3
Intravenous drug addicts	265	8.2
Spouse of AIDS pt./sero-positive	41	1.3
Others	279	8.8
Total	3183	100

AIDS is caused by human Immuno-deficiency virus (HIV), a retrovirus that attacks immune system by destroying lymphocytes Cells (T4). T4 cells are also called as CD4 cells and white blood cells. These T4 cells help in activation of immune responses. When these T4 cells are infected with HIV, it leads to breakdown of immunity in the body and cause disease, which results in death of a person.

Actually these HIV resides in the circulatory system. HIV progresses in three phases. In first phase of infection few T4 cells is infected. After some years few more cells get infected and at last, body loses its capability to replace the damaged cells as a result of which free viruses flow into the blood and damage the immune system. The first stage is over so fast that no body notices it. Such type of people get infected with flu. The second phase starts when the virus enters the lymph nodes of the body and HIV multiplies at this stage, body reactivates its defense system by producing antibodies to neutralizes HIV and seeps out of lymph nodes. At this stage another group of white blood corpuscles attack T-cells and destroys the infected cells. The third stage begins when the number of T cells drop from 1,000 cells per millimeter of blood to less than 500.

Background of AIDS

The first infected person came from serum collected in central Africa in 1959. The serological evidences of HIV were also found in N. America and Europe in 1970 and rapid spread was recorded subsequently. AIDS has produced three worldwide epidemics.

- HIV infection 21 AIDS epidemics and social cultural economic and political response to first two epidemics.
  HIV infection began in 1970 and continues till today in the form of pandemic of HIV infection.
- AIDS epidemics started in 1981 USA. The centre for disease control reported the occurrence of pneumocytes carini and Kapoosi's Sarcome in New York and Los Anglis.
- Social, cultural, economic and political epidemic is an intense spread of AIDS, which has struck the people in the age group of 20-40 years.

Table 7.6 Modes of Transmission (HIV)

Modes of Transmission	Estimated efficiency	% of total infection
Blood Transfusion	>90%	5%
Sexual intercourse	0.1 to 10%	75 to 80%
Infective drug abuse	0.5 to 1%	10%
Others	<0.5% Small	

HIV/AIDS spreads by

Spread of HIV takes place by the following ways:

- Unprotected sexual intercourse with an infected person.
- Using needles/syringes contaminated with the virus from an infected person.
- Blood transfusion from an infected person.
- From an infected mother to the foetus through the placenta.

HIV/AIDS is not spread by

- Hugging and touching.
- Sharing food and utensils.
- Shaking or holding hands.
- Mosquito bites.
- Looking after an infected person.
- Working or playing together.

- Sharing toilets.
- Coughing or sneezing.
- People in a crowd.

Symptoms of HIV

- Itchy skin rashes.
- Cold sores all over the body.
- Swollen lymph glands.
- Loss of weight (more than 10%).
- Persistent fever.
- Persistent diarrhoea.
- Severe night sweats.
- Persistent cough.

Prevention/Protection From AIDs

- Practicing safe sex.
- Avoiding injections with contaminated needles.
- Testing blood for HIV (by blood banks).
- Avoiding/terminating pregnancy if a woman is suffering from the disease.

AIDS is controlled by

- Abstinence.
- Faithfulness.
- Consistent use of condoms.
- Do not use un-sterile needles.

Treatment of AIDS

ELISA test is first technique to identify HIV.

ELISA test detects HIV antibodies in 31-60 days (Western blooding) old cases. The particle agglutination test is also screening test. Immunofluroscent assay IFA and radio-immuno precipitation assay RIPA, B-2 macroglobulin and serumneopterin are the useful tests for the diagnosis of AIDS.

Drugs used for Anti HIV are Dideoxyinosine, DDI, Zalcitabine or DDC and Zidov-udine AZT. Out of all these AZT has proved as effective drug for treatment for HIV infection. The latest drugs Ritonavir, Indianir or Crixivan and saquinavir are used at the initial stages of infection.

Status of India about HIV dispersal

In India AIDS entered in 1980. The first cases were identified in 1986-87 in Madras, Vellore and Mumbai. It was first found in a female sex worker and then spread to others sex workers, their customers, truck drivers and seasonal laborers. AIDS in India is transmitted heterosexually, like in Africa. According to Indian government 25% donated blood is untested for HIV and it is main source of transmission. Mainly there are four ways one can get AIDS:

- Sexual intercourse male-to female, female to male and male to male.
- Blood transfusion.
- Unsterilized needles.
- Sharing of injections.
- Breast feeding.

National AIDS control programme

World bank, WHO, and Government of India has launched a comprehensive scheme National AIDS control program, NACO.

Major components of the program

- To strengthen program/management capabilities through planning, implementing, counseling and monitoring the activities through AIDS control cells at the state level.
- Improvement in the upgradation of blood bank in HIV screening of all blood products.
- To strengthen the clinical services.
- To strengthen the monitoring planning and programming interventions.
- To strengthen the public awareness and community support.

HUMAN RIGHTS

The term human rights means the right relating to life, equality, liberty and dignity of the individual guaranteed by the constitution.

The Indians had hopes and aspirations of better living, true human dignity and had dreamt of economic self-reliance in 1947. People looked for an equal society. The Indian constitution enables the people to achieve aspired development in the nation. The industrial self-reliance was the solution to every problem of the country. Leaders, businessman, academicians and people were taken away by this euphoria. India was set for scientific and technical growth.

- First Human rights is to be free from government violation of the integrity of the person. The violations include cruelty, torture, degrading treatment or punishment.
- Second human right is the fulfillment of food, shelter, health care and education.
- Third human right is the right to enjoy civil and political liberties.
- According to united nation system human rights are of two types.
- Civil and political right.
- Economic, social and cultural right

Civil and Political Right

Civil rights are related to the protection of the right to life and personal liberty political right refers to the rights, which allow a person to participate in the government of a state.

Economic, Social and Cultural Rights

These rights are related to the guarantee of minimum necessities of life to human beings. Right to adequate food, clothing, housing and adequate standard of living and freedom from hunger, right to work, right to social security, right to physical and mental health, right to education. All these rights are included in the international covenant on economic social and cultural rights.

Human Rights Education

Human Rights Education is the only guarantee of demonstrating our commitment to human dignity, promotion of an adequate standard of life and finally for the promotion of world peace and prosperity for all.

Human Right Education brings awareness of fundamental human rights, freedom to all without discrimination of race, sex, and population status.

Objectives of NGO's Program for Human Right Education

A. Participants understand that they could enjoy all human right.

- They understand the conceptual meaning of human rights.
- They understand that human rights education is necessary for their socio-economic and political empowerment.
- They know the human rights spelt out in the international charter.

- They understand that discrimination should be eliminated between sex, race, languages, and nationality.

B. Participants develop positive attitude towards life and living.

- Healthy attitude towards gender-equality, neighbours and society.
- Healthy attitude towards a job.
- Healthy attitudes for the welfare of society.

C. Action plans are developed and implemented by participants. NGOs and members plan activities, which bring about social change and includes—

- Planning and implementation of activities in a phased manner, which bring social justice and economic prosperity.
- Participation in general and local elections relating to political governance of the country.
- Planning strategies to fight against and eliminate evils, abuse of power and violation of human rights.

Promotion and Protection of Human Rights by the United Nations

Following are the ways, which protect the human rights by the United Nations:

- **Human rights consciousness**
  United nation has made the people and the states conscious about the human rights and fundamental freedom.
- **Codification of the law of human rights**
  United Nation has codified the different rights and freedom by making treaties for all sections of the people.
- **Monitoring the human rights**
  A human monitor works in compliance with conventions and investigates allegations of human rights abuses. United Nations human monitors have been sent to many countries like Cambodia and worked as part of peace- keeping operations in Yugoslavia etc.
- **Compilation of information on the violation of human rights**
  United Nations compiles information on the incidents of certain kinds of violations in a specific country. A special group performs this task. This groups gathers facts and contacts with local groups

and make recommendations to government authorities, strengthening the human rights.

- **Procedure for individual complaints**
  A number of human rights treaties permit individuals to make petition before the appropriate bodies. The commission on human rights is authorized to discuss human right situations anywhere in the world.
- **Examination of human rights situation**
  The commission on human right may ask the secretary general to intervene or sent an expert to examine human rights.
- **Coordination of human rights activities**
  High commissioner is the post for strengthening the human rights activities. He protects all human rights and maintains a permanent dialogue with the member states.
- **By providing advisory services**
  The center for human rights provides advisory services to governments seeking to improve their human rights performance.

So in all, human rights serves as instrument in changing society. This also strengthens the development process and sustainable development.

ROLE OF INFORMATION TECHNOLOGY IN ENVIRONMENT AND HUMAN HEALTH

Computer is the strongest tool of information technology. Now-a- days information technology has made a great impact on every activity of mankind and plays an important role in the functioning of life. It is not only restricted to scientific and engineering applications but now much of the work is done by information technology. In the present scenario, thinking of progress without the use of information technology is just like dreaming for winning the race without the help of an expert in games. It provides friendly working, accuracy, time saving and huge money etc.

Role of Information Technology in Environment

Information technology plays important role in the new operating environment. Environmental problems are now considered global issues with trade implications. Public and consumer environmental awareness is on the rise. Information technology has made environmental information available at our finger tips. Reversed logistic and green issues are increasingly being recognized in the world trade. Manufactures are being asked to take back the used packaging and products for reuse and recycling.

Environmental professions in the third millennium will have to use the best tools available with the new age of information technology. Environmental problems are complex and ever changing. They require careful and system-solving strategies to ensure that solving one problem should not create other problems.

Applications of information technology in environment

- **Geographic information system GIS:** Geographic information system is one of the many information technologies that have transformed the ways of geographers in researches and has contributed to the society. It is an excellent tool for computerized storage and analysis of spatial data.
- **Spatial Statistics:** Statistical analysis and modeling of spatial patterns and processes are based on computer technology. Advances in information technology have made these techniques more widely accessible. They provide also more accurate information of real world processors.
- **Photogrammetry and remote sensing:** Information technology have made both photogrammetry and remote sensing readily available and far easier to use.
- **Cartography and computer assisted drafting:** Computers offer the advantages to cartographers by providing software for cartographic production.

Information Technology in Human Health

With the adventure of information technology, many hospitals have a facility to provide complete medical history in the hands of a patient. This automated system of computer has reduced the time involved in obtaining a patient's history. The patient's medical history can be printed on the medical slip and sent to the physician for review.

The information technology helps the doctors in the following way:

- To design and test pharmaceuticals
- To test and diagnose diseases and illnesses faster and in a more accurate way.

- Build and use devices to monitor vital signs and other body functions.
- To design prosthesis and reconstruction models.

Following are the techniques used in the information technology for detecting the various diseases:

1. **Multi-phasic Health Testing (MPHT):** MPHT is a series of tests with the computer aided equipment. It stores the results and reports them to doctors. This report is delivered to a doctor before patient's visit. MPHT help the doctors to spend more time on treatment and can be valuable in preventive healthcare.
2. **Magnetic Resonance Imaging (MRI):** This technique is used for getting information about the nerve tissues in a human body. The source of imaging in MRI system is a series of strong magnetic pulses applied to a body encased in a magnetically shielded chamber. Computer processing gives an image, which show relationships of bone structure and nerve soft tissues. MRI examinations do not expose patients to radiation.
3. **Prosthesis Design and Reconstruction:** Orthopedic surgery needs the creation and implantation of artificial limbs. By using diagnostic and scanning procedures, technicians can digitize an accurate picture of the damaged bone and visually compare it for fitness and function. The prosthesis is designed, with the help of computer aided manufacturing machinery, to ensure that precise measurements are met. Tooth crows can be created in the dentist office with the help of a computer.
4. **Computerized Axial Topography (CAT):** In this, computer controls the operation of X-ray equipment and also evaluates the image produced. CAT helps the doctors in re-constructive surgery and testing breast cancer.
5. **Pharmaceutical Design and Testing:** Computer techniques help in developing new drugs. It helps in determining the shape of a molecule, which shows the behaviour. It also helps in checking toxicity of a drug without endangering animal life.
6. **Computer Aided monitoring:** Computer aided monitoring is used in vital signs such as heart beat rate, blood pressure, respiration, temperature and so on. Computer-controlled devices are used where patients need constant monitoring such as in ICU (Intensive

Care Units), post-operative recovery rooms and premature baby nurseries.

Role of Women and Child Welfare

India is on the path of knowledge based, society; due to women's awareness, education etc. Knowledge in women is a great force in our country. Traditionally, women are respected in Indian society. Hence, women's economic empowerment will lead to their political empowerment. Women have a crucial role to play in projecting our culture and our way life. She plays a powerful and positive role in confidence building and creating awareness in her children and family.

The key to a rapid growth of any country lies in real economic growth and investment in its children. Care for children must be the first concern of any country for better development. Environment, education and health are the first all scale welfare programme for the children and women in a country. In India Reproductive and Child Health (RCH) approach was launched in October 1997. RCH covers all aspects of universal immunization program (UIP), child survival and safe motherhood (CSSM) programme.

As India is progressing towards commercialization and industrialization, it is becoming environmentally unsustainable. The livelihood of rural communities is badly hurt by environmental disasters, which are as follows:

- **Water pollution:** Water pollution occurs as a result of sewage, garbage disposal, industrial waste, pesticides and sedimentation through soil erosion and timber logging.
- **Toxic Wastes:** Toxic wastes are dumped in rural areas and impose severe penalties on women.
- **Deforestation:** Deforestation occurs through timber logging resulting in the loss of topsoil landslides and floods.

It is well documented that massive deforestation has serious consequences on rainfall and climate patterns, resulting in the loss of crops. These food shortages and price increase affects the livelihood of families and increase women's burden as a resource manager who have to seek out increasingly scarce alternatives. The increasing degradation of the natural resource base in many countries has thereby undermined women's role as major resource users and managers in providing food and securing overall family welfare.

Forests are the source of food, fuel and medicines. They are also the home of indigenous communities. Deforestation has caused the loss of these essentials for many indigenous, as well as rural communities. Women as resource managers now need cash that can be earned only through employment. But in a situation of gender asymmetry, even among the poor, women are the poorest of the poor with little access to scarce resources, including employment opportunities. Pushed into "Formal employment", rural women usually enter the labour market as unskilled, low-wage workers. Even so, they still carry their traditional burden of child bearing, rearing and overseeing family welfare. Thus children often constitute women's only economic wealth and social security. In the absence of viable alternatives, poor women are motivated to maximize their "wealth" by having many children.

Women of childbearing age are more vulnerable to environment pollutants then men because of their reproductive function and breast feeding practice. With the increase in number of chemicals like lead, the impact on women and children appear to be increasing. According to United Nations studies, the possible effect of exposure to chemicals results in infertility, miscarriage, malformation and neonatal death to growth retardation. Maternal and infant mortality rates are thus a significant indication of health conditions, affected not only by poverty but also by environmental pollution.

Role of women and child welfare has changed very much with the advancement of the technology in agriculture. India is an agricultural country as most of the women are dependent on the agriculture. The case study shows the marginalisation of rural woman in agricultural production as a result of technology change. The changes in the rice technology of agriculture are:

- A shift in type of seed from the traditional varieties to the short, early maturing one, produced by the international rice research institute.
- The spread of farm mechanization in the land preparation and thrashing task.
- Changes in certain farm practices such as extensive use of fertilizers, pesticides, weedicides and methods of direct seeding.

These technological changes displaced more women than men. Under the old technology almost half of the labour working per hectare of land were women. The fact that displacement occurred in the absence of other job opportunities in these areas of land. The worsening unemployment and displacement of women in other villages and towns increased competition for the marginal jobs. The collection and processing of left over grain symbolized relegation of woman to the left over employment opportunities of men. This case study also illustrates ability of economically marginalized rural women to make something out of nothing.

Women have the ability to convert waste into a food resource for their families. The economic margnialization of women nevertheless has repercussions on the welfare of both women themselves and their children. The family is the responsibility of women. The burden of restructuring the budget and meeting cash shortages seems to fall on women.

REVIEW QUESTIONS

Short Type Questions

1. What is population growth?
2. Name types of population growth.
3. What is population explosion?
4. What is mortality?
5. What is Fertility?
6. What do you mean by environment?
7. What do you mean by health?
8. Define value.
9. What is value education?
10. Define AIDS.
11. Give the full form of AIDS and HIV.
12. What are the major program involved for controlling AIDS?
13. What is Human right?
14. What is information technology?

Descriptive Questions

15. How does population growth show variations among various nations?
16. What are the preventive measures for population explosion?
17. Discuss the methods for family planning.
18. What do you mean by environment and health and how can we prevent our health by reducing environment degradation?
19. Discuss the components of value education.

20. What is the importance of value education?
21. What are the sources of AIDS?
22. How AIDS spread and what are the methods to control the disease?
23. What are the objectives of Human right?
24. What are the steps taken for promotion and protection of human rights by the united nation?
25. Discuss the role of information technology in environment and human health.
26. Discuss the role of women and child welfare in relation to environment with suitable examples.

8

Field Work

CHAPTER OBJECTIVES

- Visit to a Polluted Site

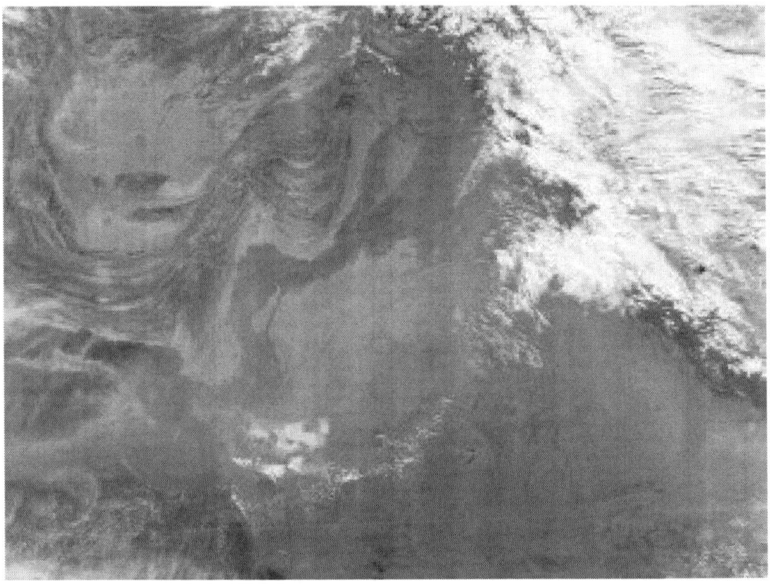

VISIT TO A POLLUTED SITE

Any undesirable change in the quality of air, water and soil that affects the health of an organism is called Pollution. Mainly pollutants are unwanted solid, gaseous and liquid chemicals that come from wastes during use, process and extraction. Pollution is also a form of unwanted energy emitted in the form of heat, radiations or noise.

Some pollutants come from a single point while some come from unidentified large number of points. These pollutants are biodegradable as well as non biodegradable in nature.

Experiment No. 1: To visit an urban polluted site.
Requirement: Notebook and pen.
Observations:
Principle: Urban areas are not self sustaining. They survive by importing food, water energy and other resources. As a result they produce a large amount of wastes that can pollute the air, water and land.

Table 8.1 Survey of Polluted Sites

S.No.	Polluted sites	Observations
1.	Industrial area	Foul smell, dizziness, large amount of industrial effluent
2.	Ditches	Evolution of gas, foul smell, infectious organisms
3.	Sewage leakage	Foul smell
4.	Heaps of solid wastes	Spread of infectious organisms
5.	Transportation	Large amount of air pollution
6.	Uneven road sites	Causes air pollution
7.	Dissolution of building and materials	Solid waste

Experiment No. 2: To visit a rural polluted site.
Rural area is with cropland and well drained fertile soil. But due to mismanagement and human activities, area became polluted.
Requirement: Note book and pen.

Table 8.2 Observation of rural polluted site

S.No.	Rural polluted site	Observations
1.	Uneven streets	Filled with huge amount of polluted water
2.	Solid wastes	Heaps of solid wastes at a very short distances
3.	Ponds	Water is polluted due to washing, bathing of animals and human beings
4.	Animal dungs	Lying at short distances in open
5.	Holes of water	Small amount of stagnant water lying in the small holes causes infectious disease

Experiment No. 3: To visit an industrial site
Requirement: Note book and pen.

Table 8.3 Survey of Industrial Site

S.No.	Activity	Observations
1.	Dumping of solid wastes	Produce large amount of foul smell
2.	Disposal of industrial effluent	Disposed off without recycling that affect the nearby land area
3.	Removal of CO_2 gas with or without filters	Emission of CO_2 and other gases without the use of filters. A dark zone is created in the atmosphere
4.	Industrial act	Industrial act is not followed by the owner of industry
5.	High stacks of chimney	Install very long stacks of chimney
6.	Waste treatment plant	Treatment plant is installed but not in working order
7.	Plantation	Plantation is not done
8.	Medical facilities for labour	Very good medical facilities provided to labour

Experiment No. 4: To study common insects, plants and birds in your surroundings

Table 8.4 Survey of Environmental Habitat

S.No.	Name of plant	Area	Utility
1.	Neem	Near house	Good insecticide
2.	Pipal	Near shopping centre	Good air filter
3.	Banyan	On the roadside	Air purifier
4.	Mango	On the roadside	Nutritious value
5.	Eucalyptus	On the roadside	Consumes more water

Table 8.5

S.No.	Name of Insects	Area	Disease Caused
1.	House Fly	House	Causes diseases
2.	Mosquitoes	Open area	Causes diseases
3.	Cockroach	Sewage water	Causes diseases
4.	Earthworm	Soil	Improve fertility
5.	Butterfly	Garden	Suck nectar
6.	Small insects or zooplanktons	Ditches	Causes disease
7.	Ant	Surroundings	Causes pain

Table 8.6

S.No.	Name of Birds	Area	Mode of feeding
1.	Peacock	Near lake	Sweet voice (omnivorus)
2.	Vulture	Near dead bodies	Feeds on dead bodies
3.	Crow	Surroundings	Feeds on solid wastes
4.	Sparrow	Surroundings	Feeds on grains
5.	Parrot	Zoo	Feeds on grains

VISIT TO A FOREST

1. Forest cover 34 per cent of the world land area. Forests are home to 50–90% of Earth species.

Experiment No. 5: Discuss the type of forests, number and ecological niches of the given area.

Table 8.7 Survey of a Forest Area

S.No.	Types of Forests	Number	Ecological Niches
1.	Old growth forest	4	Dead trees (Snag), Fallen logs (Boles)
2.	Second growth forest	10	Fallen leaves

Old growth forests are the uncut forest.
Second growth forest are the stands of trees resulting after cutting.

Experiment no. 6: Write the parts of trees and their uses.

Table 8.8 Uses of Plant Parts

S.No.	Plant parts	Uses
1.	Foliage	Oils, decorations
2.	Gums	Chewing gum, ointments, perfumes, resins, drugs
3.	Sap	Sugar & Syrup
4.	Cordwood	Fuel, charcoal, tanin, dyes, wood alcohol
5.	Bark	Tanin, drugs, oil, dye
6.	Logs	Umber, ties, veneer, plywood, press board
7.	Stumps	Veneer, Turpentine, Pine oil, Wood tar
8.	Roots	Smoking pipes, tea & oil

Experiment no. 7: To study the river as an environmental asset.

Table 8.9 Survey of a river

S.No.	Parameters	Observations
	Name of river	**Local**
1.	Flow rate	Slow
2.	Temperature	Normal
3.	Entry of pollutants	Very small
4.	Size	Small

Table 8.9 Contd...

S.No.	Parameters	Observations
5.	Water colour	Transparent
6.	Presence of water	Throughout the year
7.	Use of river water	For irrigating fields only
8.	Impact on the surrounding environment	Lowers the surrounding temperature
9.	Type of vegetation in the surroundings of river	Euclayptus
10.	Any industrial unit is present in the nearby area	No
11.	Washing of clothes	Yes
12.	Bathing of human beings	Yes

Visit to a Grassland Ecosystem

Experiment No. 8: To determine the total no. of individuals of a grass species in a given grassland system.

Requirement: Nail, Thread, Book and Pen.

The experiment is conducted to determine the diversity of grass species.

Method: Put the nails at the four corners of a grass area. Then count and note the grass species in each area covered by four nails.

Table 8.10 Survey of grassland ecosystem

S.No.	Species	No. of individuals in each area covered by four nails 1 2 3 4 5 6 7 8	Total no. of individuals
1.	Cynodon sp.	– 1 – 2 – 2 – 3	8
2.	Dactylon sp.	2 – 2 – 1 1 – 2	8

The analysis of the above data shows presence and absence of a particular grass species in a specific area as well total number of individuals.

Visit to a Mountain

Experiment No. 9: To study the mountain area

Table 8.11 Survey of mountainous habitat

S.No.	Parameters	Observations
1.	Name of the Hill	Aravali in Faridabad
2.	Type of Vegetation	Grasses, small trees
3.	Profession of local people	Agriculture and jobs
4.	Tourism	Good
5.	Population	Densly populated
6.	Means of transportation	Buses, Train, Cars

Experiment 10: Name the hot spots of your city

Hot spots area of city:

1. Lake
2. Near river
3. Corners of city
4. Parks
5. Zoos

Experiment no. 11: What are the factors responsible for degradation of trees in a given area.

Table 8.12 Survey of harmful factors for trees

S.No.	Factors	No. of trees affected
1.	Industries	5
2.	Sewage effluent	4
3.	Water logging	2
4.	Saltation	6
5.	Transportation	2

Appendix

1

Degradation of Marine Water

The International Oceanographic Commission (IOC) for the United Nations Educational and Scientific Commission (UNESCO) defines marine pollution as "Introduction by man, directly or indirectly, of substances into the marine environment (including estuaries) resulting in such deleterious effects as harm to living resources, hazards to human health, or hindrance to marine activities and reduction of amenities."

About 71 per cent of earth surface is covered by oceans, which have a volume of 1.37 × 1039 litres, weighing about 1.40 × 1036 metric tons. Rivers discharge 2 × 1014 metric tons of water into the oceans annually. Each year 4.5 × 1011 metric tons of dissolved materials is also carried by this water flow. Some of the material comes from the atmosphere with the falling rain. The amount from atmosphere is more as compared to the material which is carried by rivers.

The point of concern is that this material is not uniformly spread along the whole oceanic volume, it is accumulated near the source of pollution. Coasts are dumping grounds, having a much higher pollution concentration not only because the material is being put into these relatively shallow areas much more rapidly than it is being carried away by natural wave motions but also because the normal structure of the oceans tend to prevent the mixing of these inputs with the rest of the oceanic volume.

Marine pollution may also be of natural origin. For example, it may be caused by underwater gas and oil eruptions, or by the intensive growth and subsequent death of certain micro-organisms, which are familiar disastrous 'red tides' or by volcanic eruptions. But, such processes and effects are not regarded as pollution.

Marine pollution can be classified into three categories from ecological point of view

1. Biogeochemistry of pollutants
2. Marine ecotoxicology.
3. Biological principles of anti pollution measures.

First category includes the investigation of sources of pollutants, mechanism of their entrance in marine environment, their bioaccumulation, transformation and other processes which determine the fate of toxicants in the sea. First, the amount of toxicant is measured in the marine organisms and communities, this is quite tedious task due to low level of pollutant.

'Marine ecotoxicology' involves the study of biological effects and consequences of human interference with the composition of marine environment. These investigations result in scientific foundation on which is to be based national or international efforts to preserve the seas from pollution.

Solutions to the problem of marine pollution must be based on the results of closely interlinked ecotoxicological and biogeo-chemical studies.

MARINE POLLUTANTS

A list of common pollutants present in the marine environment have been given in the table 1.1. Their amount can vary from site to site. This difference in amount comes from local conditions like population pressure, flushing and climate.

Table 1.1 Pollutants present in the marine environment

Sediments	Petroleum
Heat	Radioactive materials
Brine	Acids and bases
Pathogens	Toxic organics
Solid wastes	Nutrients
Fresh water	Oxygen Demand Materials
Toxic inorganics	Aesthetically displeasing materials

The global pollution zone of the marine environment may be characterised as follows: highest pollution levels in the euphotic layer, high pollution levels in the neritic zone and inland seas, latitudinal distribution of toxic pollutants, a mosaic distribution pattern of toxicant concentrations in water, localisation of toxicants in the hyponeuston and benthos biotope; and a coincidence of the maximum pollution zones with zones of high biomass and high productivity.

Some general conclusions about marine pollutants are

1. The rate of biological accumulation and exchange of micro nutrients of metals and radionuclides in the world oceans is directly correlated with their physicho chemical properties and with the surface area of marine organism.
2. Super saturation of surface water by calcium carbonate largely determine the forms of micro components in sea water.
3. With increasing atomic number of the metal, the concentration of metal in sea water and in plankton tend to decrease.
4. At lower trophic levels, bioaccumulation results in bio sedimation and self-purification of surface water takes place.
5. The rate of degradation of petroleum and chloro-organic toxicants in the marine environment is less as compared to their input. The degradation process is quite slow at greater depths and the latter must be considered as a depot of toxic substances and their metabolites. The distribution of petroleum and chlorinated hydrocarbons in the world ocean in biotic and abiotic components of the same ecosystem are altogether different.
6. Generally, there is a direct relationship between the concentration level of toxicants in commercial animals and the pollution level of the marine environment.
7. The inhibitory effect of pollutants increases in the following sequence: oil, detergents, metals (arsenic, lead, cadmium, copper, mercury, methyl mercury) and chloro organic compounds. The most toxic compounds are DDT, polychlorinated biphenyls, mercury and methyl mercury which inhibits photosynthesis in long term experiments at concentrations as low as 1–10 μg/l.
8. The varying response of different monocultured species of toxic impurities in short term experiments are probably due to the different rates of absorption and penetration of toxicants in cells.

PATHOGENS

These are the living organisms that can produce sickness in either animals or plants within the ocean itself or to the humans who come in contact with oceanic water or the food from ocean. These are virus, bacteria, protozoa, fungi and so on. The most common of these are present in sewage.

SEDIMENTS

These are already present to some extent in the ocean water. These affect the plants by blocking a large amount of light, thereby decreasing the photosynthetic productivity. In some cases, rooted plants get completely destroyed by this process. After deposition these cover up bottom dwelling organisms. The deposited sediments tend to make the water shallower so that, in regions traversed by ships, periodic dredging is needed.

SOLID WASTE

As the area for the dumping of solid waste is becoming lesser day by day, the use of oceans as dumping ground has increased. It is mainly sludge material leftover as a byproduct of domestic sewage treatment. If the product discharged at sea contains materials which may be leached into the oceanic environment, a serious problem could arise especially when the materials are toxic. However, when the material is inactive, a little forethought may result in some benefit to the oceanic environment. Artificial fishing reefs, for example, have been found to be very successful in some regions, serving to enhance the habitat area, especially for small fish.

EXCESS HEAT

Excess heat, results in increase in temp and other changes which may be detrimental to the organisms present. The amount of heat that is detrimental and extent of degradation is estimated by a number of factors. The various sources of this heat are fossil fueled power plant, nuclear power plant and so on, which leads to thermal pollution.

FRESH WATER AND BRINE

Fresh water may be in great demand ashore, but in the ocean it will obviously produce a marked environmental change within a small area. This may be due to poorly designed storm drainage systems, water diversion networks associated with dams and effluent of some industrial processes and so on.

The introduction of brine also leads to a number of problems due to high salinity. Fishes face osmotic problems.

TOXICS

These are the materials commonly used in industry. Some are relatively harmless, but in higher quantity they may be destructive.

The toxic organics have been the most disturbing of the modern day chemicals commonly discharged either purposely or accidently into the marine environment. These are pesticides, insecticides, herbicidesd and so on. The most disturbing toxic organics have been the pesticide such as DDT and Kepone which have the unfortunate characteristic of being more soluble in oil that in water so that they tend to collect within the fatty tissues of marine organisms.

PETROLEUM

Although petroleum may be classified as a toxic organic, it is a naturally occurring material and is biodegradable, given enough time. It enters into the marine environment due to accidents such as tanker damage or transfer loss, natural seepage, offshore production losses, losses associated with refineries, from runoff originating as drippings or disposal of used automobiles lubricants, and unburnt hydrocarbons emitted into the atmosphere as internal combustion exhaust.

NUTRIENTS

Nutrients are those chemicals which are required by plants. The activities of man have added to the total nutrient load of almost all coastal areas. When the level of nutrients become very high, plants grow unchecked so that decaying plants exist in such great numbers that the oxygen supply becomes rapidly depleted. The main sources of these nutrients are domestic sewage effluents, agricultural runoff and runoff from urban areas and so on.

RADIOACTIVITY

Radioactive materials are discharged in the oceans by nuclear power plants, nuclear power plant fuel production and reprocessing plants and other activities relating to the radio active elements. During coal combustion, radioactive elements escape into atmosphere from where these enter into the sea by rainfall. Other sources include natural radioactivity, weapons testing, mine drainage, accidental spillage and a few isolated industries.

OXYGEN DEMAND

Oxygen demand materials are those which need oxygen for decomposition. This process results in scarcity of oxygen which would normally be utilised by marine animals. Sewage sludge and the organic material produced during eutrophication due to excessive supply of nutrients are examples of oxygen demanding materials.

ACIDS AND BASES

The acids and bases result in imbalance in marine ecosystem. The oceanic pH has been some where around 8.0, slightly basic. This is maintained by the carbonate system. If the pH fluctuate from this normal value, then this results in detrimental effects on marine biota. Most toxic materials increase their toxicity under conditions of low pH.

AESTHETIC CONSIDERATIONS

These are the materials which make the ocean unpleasant to look. Tarballs, floatables, Hydrogen sulphide producing materials and coloring agents and so on.

MANGAGEMENT PROBLEMS

Although the major pollutants have been confirmed but still the management of marine pollution is quite complex. This is so because the marine environment is used by men in different ways and this multi use is having a number of ramifications. The activities of men in the ocean may produce a whole range of pollutants. Two or more pollutants may have synergistic effect to cause more harm. For example pesticides are more soluble in oil as compare to water, so, if oil is also present in the oceans then the pesticide will concentrate in the oil and it will lead to deleterious effects.

SOURCES OF MARINE POLLUTANTS

Various activities involved in the production of marine pollutants can be grouped as follows:

Pollution causing activities

1. Recreation
2. Sewage treatment
3. Industry
4. Marine commerce
5. Electrical power generation
6. Other non-industrial wastes
7. Construction

INDUSTRY

Most industries need water for their operations. In the form of effluent, this water has heat or some chemical added to it. These chemicals may be toxic substances or simply coloring agents. Waste heat is mainly generated from thermal power plants and nuclear power plants. On some sites, this water is cooled with the help of heat exchangers which have copper tubes. From these copper tubes, copper is leached into the environment which leads to bio magnification and results in harmful effects.

MARINE COMMERCE

For marine commerce, ships require a deep enough water so that they can to and from on their appointed rounds. This requires continuous dredging and then, the problem arises for the management to treat this dredged material.

Another major problem arises from oil spills which causes a great damage to biotic system.

SEWAGE TREATMENT SYSTEMS

Sewage treated effluent has considerably high amount of nutrients which leads to eutrophication. The effluent also comprises heavy metal ions which accumulate in the biotic system and has detrimental effects.

AGRICULTURAL SOURCES

The largest single class of non-industrial polluters are farms. Agricultural pollutants have been commercial fertilisers, animal wastes, pesticides, herbicides and sediments. These ultimately lead to bioaccumulation in biota.

RECREATIONAL ACTIVITIES

In shallow water motor boats stir up the sediment on the bottom, a condition which persists for long period of time. This turbidity influences natural plant growth.

In our world there is alternative of everything except water. We can use solar energy instead of electrical energy but nothing can be used in place of water. Water has unique properties. Nothing in this world can replace water hence we should judiciously use our precious natural resource.

Appendix

2

EARTH: A UNIQUE PLANET

The earth has several parts that play important role in supporting life. The life is found in the parts of atmosphere, lithosphere and hydrosphere. Living organism make up the biosphere or ecosphere. Life on earth exists in a thin film of air, water and rock in zone extending about 60 meters below the ocean surface to about 6,000 meters above sea level. The zones of earth are discussed below.

Atmosphere

The atmosphere is a mixture of different gases, particles and aerosols collectively known as air which envelops the earth. The atmosphere provides various functions, least is it's the ability to sustain life. The atmosphere protects us by filtering out deadly cosmic rays, powerful ultraviolet rays from the sun, and even meteors on collision course with earth. Although traces of atmospheric gases have been detected well out into space, 99.1 of the mass of atmosphere lies below 25 to 30 km. Altitude, whilest 50 per cent is concentrated in lowest five km (less than the height of mountain everest).

Air: Air remains remarkably uniform in composition and is the result of efficient recycling process and turbulant mixing in the atmosphere. Such recycling and mixing of air help to minimise the amount of time which man made pollution spends in the atmosphere at any single location, thereby reducing the environmental impacts. The gases present in atmosphere are Nitrogen, Oxygen, Carbon Dioxides, Trace gases and so on.

Table 2.1 Relative proportion of gases (By Volume) in atmosphere

S.No.	Gases	Symbol	Percent
1.	Nitrogen	N_2	78.084
2.	Oxygen	O_2	20.949
3.	Argon	Ar	0.934
4.	Carto dioxide	CO_2	0.0318
5.	Neon	Ne	0.0018
6.	Helium	He	0.0005
7.	Cryption	Kr	0.00011
8.	Xenon	Xe	0.00009
9.	Hydrogen	H_2	0.00006
10.	Methane	CH_4	0.00021
11.	Nitrogen Sulphur	NOx SOx	0.0005
12.	Ozone	O_3	0.000004
13.	Water vapour	H_2O	variable Concentration

In reality, however the atmosphere contains a large amount of water vapour, resulting from evaporation from surface water and transpiration from plants. Near the surface this is typically about four per cent (by volume) but only 3–6 parts per million (ppm) above 10 km.

Nitrogen: The gas nitrogen composed of two nitrogen atoms and occupies nearly 78 per cent of earth's atmosphere. It is colourless odourless and tasteless. It is essential to the nutrition of plants and animals. Nitrogen is a constituent in proteins and genetic material. Some species of bacteria can absorb nitrogen from the air and convert into ammonia. Ammonia is used by the plants. This process is called Nitrogen fixation. Members of family leguminaceae can fix the atmospheric nitrogen with the help of bacteria living in their roots. However with lightening the atmospheric nitrogen is also absorbed by the soil.

Oxygen: Oxygen (O_2) is composed of two molecules of oxygen atoms. This gas occupies nearly 21 per cent of earth's atmosphere. Almost all plants and animals require oxygen for respiration to maintain life.

Oxygen is very reactive gas and oxides of various elements are formed by the process of oxidation. For example, when oxygen is combined with iron, it gives rise to iron oxide but its slowler process and generally known as rust formation.

Within the atmosphere oxygen molecule combines with another molecule of oxygen and forms O_3 (ozone).

Trace gases: Remaining gases are known as trace gases because these are present in small concentration. Noble gases which include Ne, Xe, He, Kr are very inert. Hydrogen is present in trace amounts and escapes from the atmosphere due to very low molecular weight. Ozone gas behaves as green house gas in troposphere and acts as a shield against UV radiation in stratosphere. During 20th century, anthropogenic pollution added new gases in the atmosphere like CfCs which destruct the ozone layer by causing ozone hole.

Atmospheric Layers

The atmosphere is not physically uniform but has significant variation in temperature and pressure with altitude which define a number of atmospheric layers. These include the troposphere, stratosphere, mesosphere and thermosphere. Different layers are explained below:

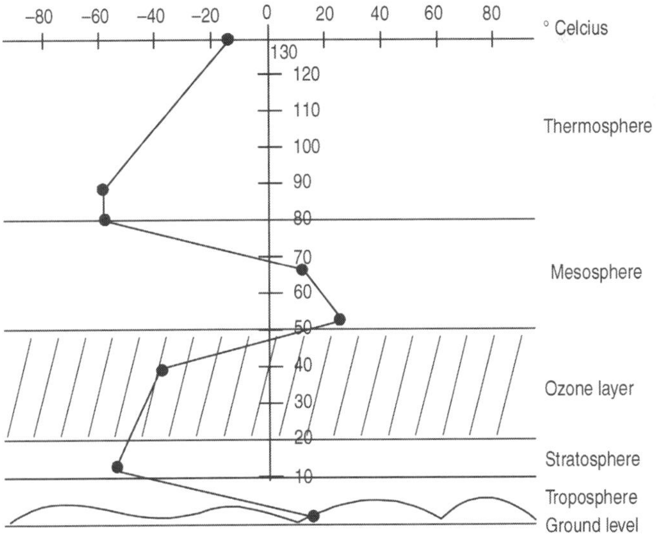

Fig. 2.1 Diagramatic representation of different layers of atmosphere

1. **Troposphere:** The lowest layer of atmosphere is called troposphere. It ranges in thickness from 8 km at the poles and 16 km over the equator. Temperature usually declines with increasing altitude in the troposphere. Troposphere contains 75% mass of atmosphere all atmospheric water vapours is found in troposphere.
2. **Stratosphere:** It is the second major layer and lies above troposphere. It occupies 12 to 50 km. This is defined as the layer in which temperature rises with increasing altitude.
3. **Mesosphere:** Mesosphere is the third highest layer in our atmosphere and lies between 50 to 80 km region. The mesosphere are the coldest atmospheric layer.
4. **Thermosphere:** Thermosphere is the outer layer. With in the thermosphere temperature rises continually. The lower part of thermosphere is called as Ionosphere and ranges between 80 to 550 km. Beyond ionosphere, exosphere (outer thermosphere merges with the space).

Lithosphere

Lithosphere means rock sphere. The solid component of the earth is termed as lithosphere. It is an important component of biosphere and mainly composed of following components

1. Elements Iron, Nickel, Carbon, Silica and so on.
2. Minerals Hematite, Dolomite and so on.
3. Rocks Solid substratum on the earth. Three types of rocks are Igneous, Sedimentary and metamorphic.
4. Soil Various types of soil like sandy, loamy, alluvial and so on.
5. Microland form Valleys, sand dunes, deltas and so on.
6. Mesoland form Mountains, plains, plateaus etc. are included in mesolandform.
7. Macro and forms Continents are the parts of macroland forms.

The lithosphere accounts for about 29 per cent of the total surface area of the globe. It provides a variety of habitat to plant and animals in the form of land mass. The geological cycles involve a set of process and sub cycles for example tectonic cycle, rock cycle, geochemical cycle, water cycle responsible for certain maintenance, changes or destruction of material on earth.

The structure of earth can be compared with that of onion and consist of three layers.

I. Cust

II. Mantle

III. Core

I. Crust: The outermost shell and the thinest layer. It makes up only two per cent of the pigment. Its average thickness varies from five to 40 km. Average thickness of the crust beneath the ocean is five km and the continental crust is 40 km. But the density of oceanic crust is more than the continent crust.

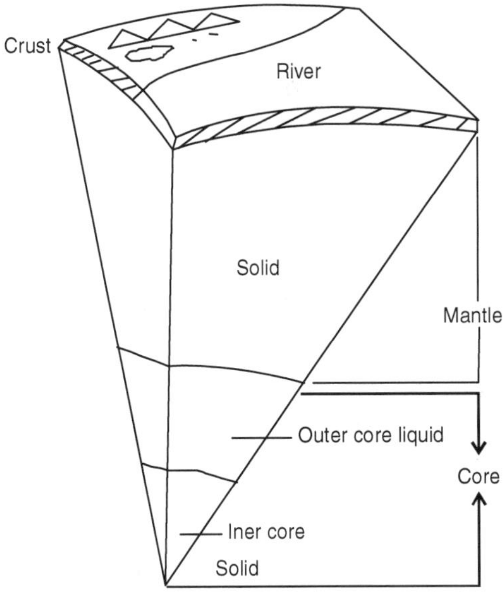

Fig. 2.2 Internal Zone of Earth

Mantle: The next layer of earth is mantle. It is thick solid zone. It begins at the depth of 10–65 km. This zones makes up 82 per cent of Earth's volume and 68 per cent of its mass.

Earth Crust Pie Chart

Core: The inner most layer of the earth is called core. It is also called as metallic core because it is mainly composed of metals. It is divided in two parts outer core which is liquid and inner core which is solid. Outer core is made up of iron and Inner core made up of iron and nickle.

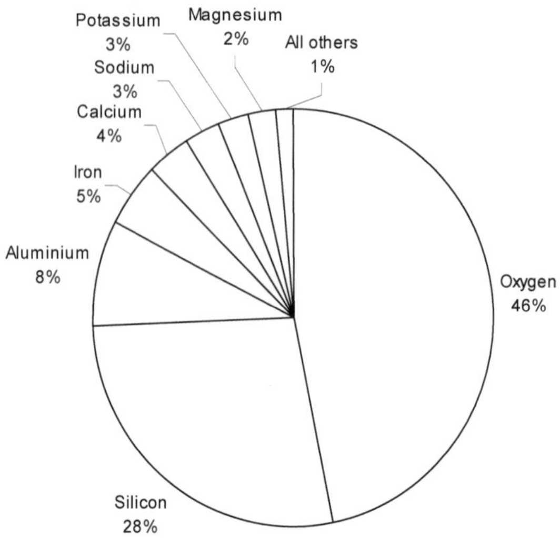

Fig. 2.3 Relative composition different elements in Earth crust

Ecosphere or Biosphere

The portion of earth where living organism are found and interact with one another and with non-living environment is known as Ecosphere. This zone of life includes most of the part of hydrosphere, upper parts of Lithosphere and lower atmosphere. Biosphere is mainly divided into two factors

1. Abiotic factor
2. Biotic factor

Abiotic factors Abiotic factors are those factors which are non-living factors and affect the biosphere by its constituents. The constituents of abiotic factor are discussed below

I. **Climatic factor:** It includes
   (a) Light
   (b) Temperature
   (c) Rainfall

II. **Physiographic factors**
   (a) Altitude
   (b) Steepness

III. **Edaphic factors**

I. **Climatic factors:** Climate of a particular region mainly depends on light, temperature and rainfall. These are discussed as below

(a) *Light:* It is the basic need for plants as well as animals. Main source of light is sun. Light helps in photosynthesis, flowering and transpiration processes of plants. Energy from the sun that is radient energy is converted into chemical energy, thermal and mechanical energy. So light affects the organisms directly or indirectly.

(b) *Temperature:* Temperature is the main climatic factor which affects the plants and animals in different ways. Temperature affects growth and development, morphology (that is, pigmentation of skin), distribution of animals and so on.

(c) *Rainfall:* Water is an important factor for life and it reaches the earth in the form of rainfall which is the result of evaporation. Various forms of precipitation are there but rain is the most important one. The amount of annual rainfall a particular area receives affects the vegetation and organisms in that region.

II. **Physiographic factors:** Physiographic factors are those physical and geographical factors which affect the ecosphere i.e., physical geography of earth affect the ecosphere. These include

1. Altitude
2. Steepness

1. *Altitude:* Mountains, valleys are formed on the earth due to different processes. With increase in height the temperature gradually or abruptly comes down and drastic change in animal and plant population is seen on mountains.

2. *Steepness:* Slope or steepness is the important feature of mountains because slope determines the water flow. Water is the main governing body life.

III. **Edaphic Factor:** Edaphic or soil factors include the soil characteristics. Soil is the most important part of ecosphere. Plants depend on soil directly and animals depend on the plant means indirectly. Soil is also important for animals.

1. *Soil system:* Soil system is very complex system and is affected by many other factors. Soil is defined as the part of earth crust which supports plant root.
2. *Biotic factors:* Biotic factors includes all living organism. Plants and animals living together influence each other. For growth and development interaction occurs within the same species (Intraspecific), between two different species (Interspecific interaction). Different types of interactions occur between the organisms. The plants are known as producers because they produce food whereas animals are known as consumers since they consume food.

Hydrosphere

The water surrounding the earth in the form of liquid water, frozen water and water vapour is called as hydrosphere.

Water is the most abundant resource on the earth, covering nearly 71 per cent of the earth's surface. Only fresh water helps to maintain earth's climate and is essential for life. Water is renewed by the water cycle. Such a cycle is hydrological cycle.

Out of total water only three per cent is fresh water and 0.5 per cent is available fresh water, out of which only 0.003 per cent fresh water is used. Major oceans of the earth are, Atlantic, Pacific, Indian, Arctic, and Antarctic. Marine water and fresh water are the types of water found. Marine water is present in oceans whereas fresh water remains in ponds, lakes and so on.

Fresh water: As said earlier fresh water occupies only three per cent of the earth as compared to marine water. Further it is divided into two parts

(i) *Standing water:* It is also called as Lentic. This included ponds, lakes, swamps.

(ii) *Running water:* It is called as Lotic system. In this water is flowing in rivers, streams and so on. Lotic system is generally divided into 3 zones

1. Littoral zone
2. Limnetic zone
3. Profundal zone

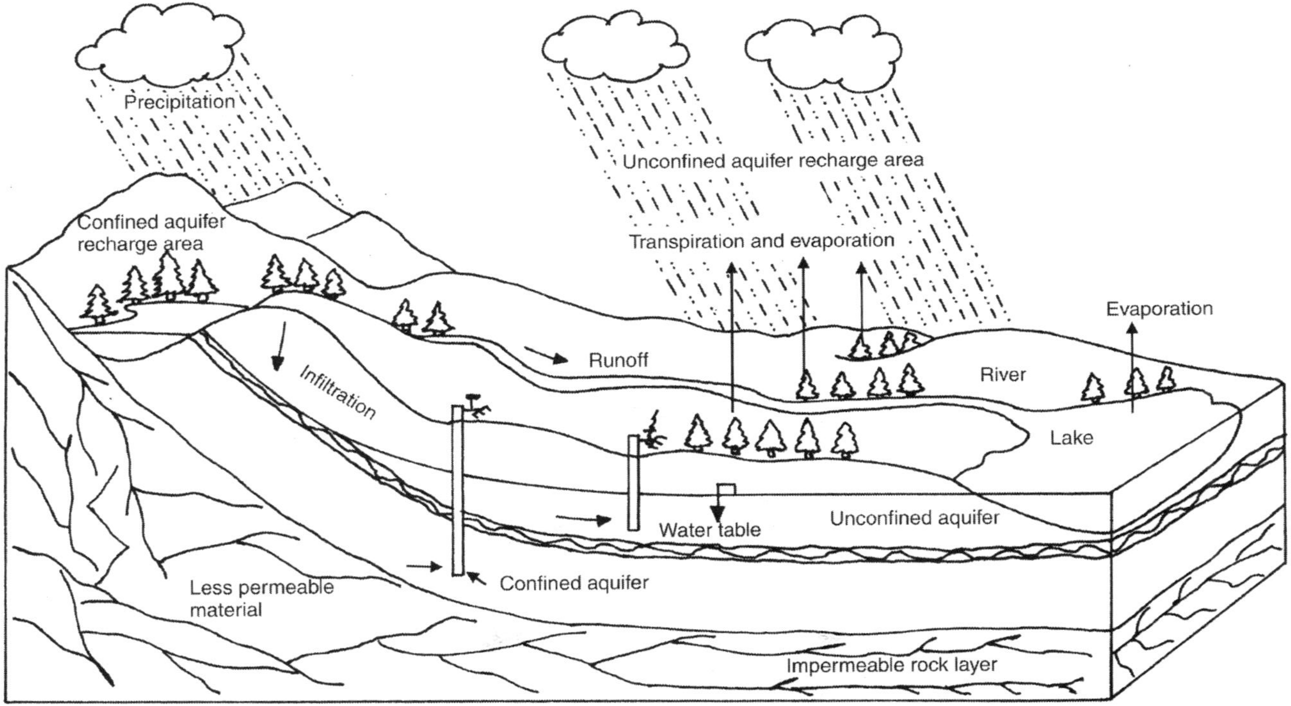

Fig. 2.4: Hydrological Cycle

Marine Water: Oceans covers 71 per cent of Earth surface. About 97 per cent water is inter-connected oceans. The average depth of the world's oceans is 3.8 km. Oceans play an important role in survival of life on earth. Oceans act as reservoir of many minerals like iron, oil, natural gas and so on. Oceans provide habitat to 230,000 species.

Appendix 3

DETRIMENTAL EFFECTS OF THERMAL ENERGY ON WATER

Thermal waste water is the warming up of an aquatic ecosystem to the point where desirable organisms are adversely affected. The accelerated pace of development, rapid industrialisation and extensive population density have increased the demand of thermal power plants. Due to human activities, adding of pollutants to the air and water are constantly increasing at an alarming rate.

Sources of thermal water

1. **Coal-fired power plants:** Some thermal power plants use coal as fuel while few plants use nuclear fuel. The condenser coils of coal-fired power plants are cooled with water from nearby lake or river and the hot water discharged into the stream thereby increasing the temperature upto about 15°C. The heated water decreases the dissolved oxygen of water. This results in death of fishes and other organisms.
2. **Nuclear power plants:** Traces of radioactive substances from hospitals, nuclear experiments and explosions, research institutes are drained into nearby water bodies. Emissions from nuclear reactors and processing installations are also responsible for increasing the temperature of water bodies. The liquid radioactive water consist of C-14, H-3, Fe-59 and Co-60 and corrossion products come during the operation of power reactors and nuclear fuel processing units. All these contribute maximum heat in the

aquatic environment. In addition, leakage of radiation from nuclear reactors in water raises the temperature of nearby water bodies.

3. **Industrial effluents:** Coal power plants, nuclear power plants, electricity generated from industries require large amount of water as compared to pulp, textile and paper industries. To cope with the tremendous demand for electricity, the generating power of the installations will be raised which will then need a large proportion of stream flow.
4. **Hydro-electric power:** Generation of hydro-electric power results in negative thermal loading in water systems. Besides electric power industries, various industries with cooling requirement contribute to thermal loading. It has been found that 18 per cent more heat is given to cooling towers in nuclear power plants than any other power plant.
5. **Domestic sewage:** Domestic sewage is mainly discharged into lakes, rivers, canals or streams with or without treatment. This domestic sewage raises the temperature of water bodies to a measurable extent but also causes, negative impact on aquatic biota. The organic matter present in the domestic sewage uses the dissolved oxygen in the water for oxidation and results in anaerobic conditions that release a foul and offensive gases in water. The organisms present in the water bodies lead to death and decay.

Negative effects of thermal power plant effluents

1. **Changes in water properties:** Due to rise in temperature, physical and chemical properties change which leads to increase in vapour pressure and decrease in viscosity of water. The decrease in viscosity, density and solubility of gases increases the settlement speed of suspended particles that affects the food supplies of aquatic fauna.
2. **Reduction in dissolved oxygen:** Dissolved oxygen concentration decreases with the increase in temperature of water. So for a healthy stream adequate supply of dissolved oxygen is required for the fauna.
3. **Interference with biological activities:** Biological activities like physiology, metabolism and biochemical processes are disturbed with the increase in temperature. Because life processes of aquatic

animals involve various chemical reactions. These reactions change with the increase in temperature.

4. **Increase in toxicity:** Toxicity increases with the increase in temperature. A 10°C increase in temperature doubles the toxic effect of potassium cyanide whereas 80°C increase in temperature triples the toxic effects of o-xylene which cause fishes death.
5. **Differences in reproductive rate:** Depositions of eggs in females is triggered by the increase in temperature. Examples are clams, oysters and shell fish. Harrying eggs hatch in 47 days at 0°C and 8 days at 14.6°C. Similarly Atlantic Salmon eggs hatch in 114 days in winter and in 90 days at 7°C.
6. **Increased vulnerability to disease:** Various pathogenic organisms activities increased by increase in temperature. Hot water causes diseases in salmon fish.
7. **Differences in metabolic rate:** Metabolic rate of aquatic organism increases. Sockeye salmon cruises swims twice as fast in water at 15°C as at 0°C, but above 15°C its speed declines.
8. **Entry of destructive organisms:** Thermal pollutants allow the invasion of destructive organisms. Entry of ship worms into New Jersey's oyster creek is a good example.
9. **Changes in algal population:** Thermal pollution causes excessive growth of algae and results in acceleration of eutrophication and disrupts the food chain.
10. **Enhances the growth of blue green algae:** Thermal discharges favours the growth of blue-green algae. This green algae causes damage to ecosystem. Because bluegreen algae are the poor food for the aquatic organisms.
11. **Demand of oxygen increased:** At higher temperature, rate of physiological process increases and favours the bacterial growth, the oxidation of oxygen demand wastes is speeded up due to high rate of oxygen depletion. Thus the demand of dissolved oxygen increases further.
12. **Longevity:** High temperature of water shortens the life of aquatic organisms. The activities of organisms is maximum between 30°C to 38°C and then decreases as the enzymes become less active.

13. **Mortality:** High temperature of thermal water increases the death rate. Only a few species can survive in a water temperature above 40°C.
14. **Distribution of organisms:** Distribution of organism is also affected in hot water. Organisms move towards lower temperature. As a result their activities are influenced by the change in surface tension, viscosity and density of water. The organisms cannot compensate for pull of gravity in a less dense water and sink down.

Measurement of thermal waste water

Following are the instruments used for measuring thermal pollution

1. Lux meter
2. Electronic temperature meter
3. Dissolved oxygen meter
4. Determination of physical parameters

1. **Lux meter:** Lux meter consists of a photosensitive detector connected to an equipment measuring the photocurrent caliberated to read in lux. Lux meter is a measure of illumination at different depths. The photosensitive detector measures the photocurrent and the extent of thermal effluents discharged in the water bodies.
2. **Electronic temperature meter:** For this purpose thermistors, thermocouples resistance thermometers and reversing thermometers are used for temperature measurement at different depths.
3. **Dissolved oxygen meter:** Dissolved oxygen concentration is an index of organic and thermal pollutants. Dissolved oxygen meter is a portable equipment. This meter is in use at Environmental Survey Laboratory ESL Rajasthan and Bombay.
4. **Determination of physical parameters:** Extent of thermal pollution can be determined by knowing the physical parameters such as temperature distribution, lux readings and dissolved oxygen concentrations.

Appendix

4

NOISE

Noise refers to sounds received by an individual. It is considered as potent and formidable pollutant and may cause noise induced hearing loss.

Decibel is the basic unit of noise. It is the ratio of two sound pressure and is equal to ten times the logarithmic ratios of pressure.

$$\text{dB} = 10 \log 10 \frac{Q}{Q_0}$$

$Q_0 \rightarrow$ is the reference standard pressure

$Q \rightarrow$ is the measured quantity of sound pressure

Human can hear frequency in the range of 20 HZ to 20 kHz. Human ear is known to be sensitive from 0 to 180 dB where 0 dB is the threshold of hearing whereas 140 dB is the threshold of pain. The ear is very sensitive at 3 kHz and has lower sensitivity at high and low frequencies. WHO has recommended 75 dB as exposure limit to industrial noise. Threshold limit value of noise universally accepted is 90 dB for 9 hrs., 100 dB for 2 hrs. and 115 dB for 15 min. per day.

Measurement of Noise

Measurement of noise can be done by sound intensity or sound pressure.

1. **Sound intensity:** Intensity of travelling wave is the energy flow per unit time per unit area to the direction of flow. Sound intensity level is given by

   1 L = 10 log I/IodB

   I is the measured intensity

I_0 is the reference intensity. Its value is 10^{-12} W/m^2.
It is directly proportional to the square of sound pressure level.

2. **Sound pressure:** It is the fluctuating part of air pressure. It is denoted by SPL. The noise level is expressed in terms of sound pressure level.

$$SPL = 20 \log 10\ P/P_0 dB$$

P = sound pressure
P_0 = reference sound pressure
and having value 2×10^{-5} N/m^2

3. **Traffic noise index:** It is represented by TNI

$$TNI = L_{90} + 4(L_{10} - L_{90}) - 30$$

L_{90} = Level exceeded 90 per cent of time
L_{10} = Level exceeded 10 per cent of time

4. **Sound power level:** It can be determined by following formulae.

$$Lw = 10\log_{10} W/Wo$$

W = power in watts of sound source
W_o = reference power level
$W_0 = 10^{-12}$ watt

5. **Noise Dose:** It is the sum of noise exposure index. Noise dose is mainly used to assess the damaging power of a day of exposure to noise level.

6. **Noise exposure index:** It is denoted by NEI. Noise exposure index is the ratio of actual exposure time at a certain noise level to permitted exposure time. It is calculated by the following formulae.

$$NEI = \frac{C}{T}$$

$C \rightarrow$ actual exposure time
$T \rightarrow$ permitted exposure time

Noise measuring instruments

Following are the noise measuring instruments used to measure noise.

1. Noise survey meter
2. Octave band analyser
3. Noise desometer
4. Sound level meter

Sound level meter is the best instrument for measuring use and easily available in the laboratories. Sound level meter is used for the evaluation of sound pressures. It indicates the rms value of the sound. In sound level meter, at microphone the sound wave energy is converted to electrical signal. The electrical signal is converted into AC signal to DC by rectifier to reflect needle to display meter to register the sound pressure level directly in dB.

Effects of Noise Pollution on Health

Exposure to high levels of noise causes harmful effects on health. These are of two types – Auditory effects and non-auditory effects.

1. Auditory effects causes hearing impairments which are as follows
   - (a) *Tinnitus*: Ringing in ears occurs in conjuction with hearing loss.
   - (b) Noise induced permanent threshold shift. It occurs on long term noise exposure. In this hearing loss is irreversible.
   - (c) *Temporary threshold shift*: It is short lived.
   - (d) *Paracusis*: Some sounds may be heard but heard incorrectly.

Non-auditory effects on health

1. Noise pollution causes decreased depth and increased frequency of respiration.
2. Noise pollution causes blood pressure reduced salivation, severe headache, nausea and fluctuations in arterial blood pressure.
3. It also causes impairment of night vision, pupillary dilation and decrease in the rate of colour perception.
4. It causes hypertension and gastrointestinal modification in high noise pollution areas.
5. It causes muscles to contract leading to nervous breakdown.

Noise pollution control in India

1. Under Environment Protection Act, 1986 and Air Act, ambient air quality standards in respect of noise of different categories of zones have been identified.
2. Motor vehicle act provides restriction on using double sirens.

So in a nutshell, sound is very necessary for human existence but it becomes hazardous when in excess. So more attention should be paid to control noise pollution. It should be handled by a large scale noise control programme. Proper noise pollution education should be provided to the noise polluted zones.

PRINCIPLES FOR UNDERSTANDING AND SUSTAINING THE EARTH

Resource Pollution and Environmental Degradation

- Resources are limited and must not be wasted (Principle of limits)
- Reducing resources consumption and waste production is the most important priority, followed by reusing items, and then recycling key mineral resources (three R'S of earth's core).
- Humans are the only species that deliberately waste resources (wastes are potential resource).
- Stress upon the use of perpetual and renewable resources (principle of sustainable yield).
- Everyone is downstream from everybody (principle of global commons).

Scientific Principles

- We cannot create matter, we can only change from one form to another (law of conservation of matter).
- Organised and concentrated matter is high quality matter that can usually be converted into useful resource at an affordable cost. Disorganised matter is low quality matter which often costs more (Principle of matter quality).
- Energy can neither be created nor destroyed. It can be changed from one form to another (first law of energy).
- Organised energy is high quality energy and can be used to do things (energy quality principle).
- In conversion of energy small amount of energy is waste (second law of energy).
- Life depends on the one way flow of high quality energy from Sun through the producers to consumers and back into space in the form of heat (Principle of energy flow).
- No population can keep growing indefinitely (Principle of carrying capacity).
- Every species has a specific role to play in nature (ecological niche principle).
- Too much or too little chemical or physical factor can limit the growth of population. (Principle of limiting factor).

- Survival of the fittest (Principle of natural selection).
- Biological communities can adapt to changes in environmental condition by gradually developing different and more complex communities (Principle of ecological succession).
- Over billions of years, nature has evolved a variety of species, (species diversity) genetic variety within species (genetic diversity), natural system (ecosystem diversity) in response to environmental change and process is carried on. (Principle of biodiversity).
- Everything is connected with everything else (Second Law of Ecology)
- Nature is more complex than we think (Principle of Complexity)
- Think globally, act locally (Principle of Change)
- We should leave the earth in as good condition as we found it (Right of Unborn Principle)
- Do not perform such acts which degrade the physical chemical and biological aspect of earth (Balanced Earth Budget Principle)

Environmental Awareness Levels

1. **Ist Level of Awareness**
   Environmental Degradation
   Environmental degradation mainly arises from the pollution which is responsible for threatening the life on the earth. Environmental problem can be solved in isolation and dealing it by use of legal, technological, and economic methods.
2. **IInd level of Awareness**
   Population Explosion
   Environmental degradation is the result of population explosion in poor countries and consumption of more energy in rich countries. Stabilise and then reduce population. Reduce waste energy and matter especially in rich countries.
3. **IIIrd Level of Awareness**
   Sustainable Earth
   Sustainable development can only take place when sustainable earth is present. This can be possible when we are working with the earth in a natural way.

How to Save Water and Money

- In new house install water saving toilets or composting toilets.
- Flush toilet when necessary.
- Take short showers.
- Do not keep water running during brushing teeth, shaving and washing.
- Compost your food. Instead of putting it in the water systems.
- Check for leaks frequentally and repair them.
- Wash the car with bucket instead of pipe.
- Use drip irrigation in garden.
- Use mulch to reduce evaporation and irrigation efficiency.
- Rain water should be harvested by making tanks.

How to Save Energy and Money

Transportation

- Ride on bicycle for short trips.
- Use bus or trains for long trips.
- Drive a car that gets at least 15 km/l mileage.
- Keep engine tuned and replace airfilter regularly.
- Keep tire pressure at recommended level.
- Consolidate trips to accomplish several purposes.

Cooking, Refrigeration and Other Appliances

- Use a gas stove instead of electric stove.
- Buy only most energy efficient store and other electrical appliances.
- Do not use electric clothes dryer. Instead use air or sun to dry clothes.
- Do not locate refrigerator near heating instruments.
- Use solar devices such as solar cooker, solar heater, solar batteries and so on instead of electrical appliances.

Hot Water Heating

- Install the most energy efficient system available such as active or passive solar heaters.
- Insulate hot water pipes and water heater.
- Do not use electrical water heater.
- Drain two quarts of water from bottom of tank every two months to reduce sediments build up and improve efficiency.
- Use low flow shower and not let water run while bathing.

Cooling, Air Conditioning and Lighting

- Buy most energy efficient air conditioning available and evaluate it only in terms of life time cost.
- Use small floor fans, ceiling fan to reduce air conditioning needs.
- Close windows and drapes on sunny days and open them on cool days.
- Turn off lights or electrical appliances when not in use.

Important Abbreviations

ASSINSEL	:	International Association of plant breeders for the protection of plant varieties
BAG	:	Basmati action group
BNHS	:	Bombay Natural History Society
BSI	:	Botanical Survey of India
CBD	:	Convention on Biological Diversity
CCC	:	Customs Co-operation Council
CGIAR	:	Consultative Group of International Research
CI	:	Conservation International (Washington, DC, USA) Earth Watch (Water town, USA)
CITES	:	Convention on International Trade in Endangered Species of Wild Fauna and Flora
CMC	:	Centre for Marine Conservation (Washington DC, USA)
CoP	:	Conferences of the Parties
CSIR	:	Council of Scientific and Industrial Research
EDF	:	Environmental Defence Fund (New York, USA)
EIA	:	Environmental Impact Assessment
EPA	:	Environmental Protection Agency
FAO	:	Food and Agriculture Organisation of United Nations
GATT	:	General Agreement on Tariffs and Trade
GEF	:	Global Environmental Facility
GIS	:	Geographic Information System
Green Peace USA	:	(Washington DC, USA)
ICEL	:	International Council of Environmental Law (Bonn, Germany)
IIED	:	International Institute for Environment and Development
IPRs	:	Intellectual Property Rights

IRRI	:	International Rice Research Institute
IUBS	:	International Union of Biological Sciences
IUCN	:	International Union for Conservation of Nature and Natural Resources (Gland, Switzerland)
IWC	:	International Whaling Commission
MAB	:	Man and Biosphere Programme UNESCO
NBPGR	:	National Bureau of Plant Genetic Resource
NPGS	:	National Plant Germ Plasm System
NRC	:	National Research Council
PBR	:	Plant Breeder's Rights
TRAFFIC	:	Trade Record Analysis of Flora and Fauna in Commerce
UNCED	:	United Nations Conference on Environment and Development
UNCTAD	:	United Nations Conference on Trade and Development
UNDP	:	United Nations Development Programme (New York, USA)
UNEP	:	United Nations Environment Programme (Nairobi, Kenya and Washington, DC, USA)
UNESCO	:	United Nations Education, Scientific and Cultural Organisation, Paris
WCED	:	World Commission on Environment and Development
WCI	:	Wild Life Conservation International (Born X, USA)
WCMC	:	World Conservation Monitoring Centre (Cambridge, UK)
WCS	:	World Conservation Strategy
WCU	:	World Conservation Union
WRI	:	World Resources Institute (Washinton DC, USA)
WWF	:	World Wide Fund for Nature

Suggested Readings

1. Reed Wicander and James S. Monroe 2002. Essentials of Geology Thomson Learning, Wordsworth group publisher California.
2. P.S. Verma and V.K. Agarwal 1983. Principles of Ecology, New Delhi.
3. G. Tyler Miller Jr. 1994. Living in the Environment Wordsworth Publisher, California.
4. G. Tyler Miller Jr. 1991. Environmental Science Wordsworth Publisher, California.
5. P.R. Trivedi 2004. Natural Resources Conservation, Delhi.
6. Krebs, C.J. 1985. Ecolony 3rd Ed. Harper & Row, New York.
7. Odum, E.P. 1983. Basic Ecology Hold-Seunder Intl. Ed. Japan
8. Odum, E.P. 1971. Fundamentals of Ecology 3rd Ed. W.B. Saunders & Co., Philadelphia.
9. deshmuch, Ian 1986. Ecology and Tropical Biology, Blackwell, Oxford.
10. Clarke, G.L. 1954 Elements of Ecology, John Wiley and Sons Inc. New York
11. K.G. Agarwal, 2002 Global Biodiversity, New Delhi.
12. Kormondy, E.J. 1969 Concepts of Ecology, Prentice Hall, New Jersey.
13. Strahlar A.N. and Strahlar A.H. 1974 Introduction to Environmental Science, Hamilton Publishing Co., Santa Barbara, California.
14. Misra R. 1968, Ecology workbook Oxford and IBH, New Delhi.
15. Billings, W.D. 1964, Plants and the ecosystem, Macmillian and Co., London.

16. Etherington, J.R. 1975. Environment and Plant Ecology, John Wiley, New York.
17. Whittaker, R.H. 1975. Communities and Ecosystems, 2nd Ed., Macmillon, New York
18. Khoshoo, T.N. 1984. Environment Concerns and Strategies, Indian Environmental Society.

Metric Conversion Chart

Length	
1 kilometer (km)	= 0.6214 mile (mi)
1 meter (m)	= 1.094 yards = 3.281 feet
1 centimeter (cm)	= 0.3937 inch
1 mile (mi)	= 1.609 kilometer (km)
1 yard	= 0.9144 meter (m)
1 foot	= 0.3048 meter (m)
1 inch	= 2.54 centimeter (cm)
Area	
1 square kilometer (km^2)	= 0.386 square mile mi^2
1 square meter (m^2)	= 1.196 square yards (yd^2)
1 square centimeter	= 0.155 square inch $(in)^2$
1 square mile	= 2.59 square kilometer (km^2)
1 square foot	= 0.0029 square meter $(m)^2$
1 square inch	= 6.4516 square centimeter $(cm)^2$
Volume	
1 cubic kilometer (km^3)	= 0.24 cubic mile (mi^3)
1 cubic meter (m^3)	= 264.2 gallons
	= 35.314 cubic feet (ft^3)
1 Litre (l)	= 1.057 quarts
	= 33.815 fluid ounce
1 cubic centimeter (cm^3)	= 0.0610 cubic inch (in^3)
1 cubic mile (mi^3)	= 4.168 cubic kilometer (km^3)

1 cubic yard (yd^3)	= 0.7646 cubic meter (m^3)
1 cubic foot (ft^3)	= 0.0283 cubic meter (m^3)
1 cubic inch (in^3)	= 16.39 cubic centimeter (cm^3)

Mass

1 metric ton	= 2205 pounds
1 gram g	= 0.03527 ounce
1 pound (b)	= 0.4536 kilogram (kg)
1 ounce (oz)	= 28.35 gm (g)

Pressure

1 kilometer/square centimeter kg/cm^2

= 0.96784 atmosphere atm

= 14.2233 pounds/square inch lb/in^2

= 0.098067 bar

1 bar = 0.98692 atmosphere (atm)

= 105 pascals (Pa)

= 1.02 kilograms/square centimeter kg/cm^2

Temperature

- To change from Fehrenheit (F) to Celsius (C)

$$°C = \frac{(°F - 32°)}{1.8}$$

- To change from Celsius (C) to Fehrenheit (F)
  $°F = (C \times 1.8) + 32°$
- To change from Celsius (C) to Kelvin (I)
  $K = °C + 273.15$
- To change from Fahrenheit F to Kelvin (K)

$$K = \frac{(°F - 32°)}{1.8} + 273.15$$

Glossary

Abiotic – Non living

Abyssal zone – It is the bottom, deep and dark zone of ocean.

Acid rain – It is the falling of acids along with the rain from atmosphere to the earth.

Afforestation – It is the plantation of trees.

Agro forestry – When crops and trees are grown together.

Air pollution – When chemicals or substances are in higher concentration in the air to cause harm to humans and other living beings. Such chemicals are known as air pollutants.

Algal bloom – Excessive plant nutrients in water lead to population explosion of algae.

Anaerobic organism – Organism that live without oxygen.

Animal manure – It is an organic fertilizer, which is made up of dung and urine of animals.

Anthropogenic – Man-made.

Aqua culture – It is a technique used for growing and harvesting of fishes in ponds, lakes, estuaris and coastal lagoons.

Aquatic ecosystem – It is a water-based ecosystem.

Aquifer – It is a water saturated layer of underground rock, which can give an economically significant amount of water.

Arable land – It is the land which can be used for cultivation to grow crops.

Area strip mining – It is the cutting of deep trenches to extract minerals.

Atmosphere – Whole mass of air, which envelops the earth.

Bathyl zone – This zone is cold and dark below the euphotic zone. Only small amount of sunlight penetrates in this zone.

Benthic zone – It is the bottom zone of water body.

Bhopal disaster – It is the disaster that occurred on December 3, 1984 at Bhopal due to leakage of methyl isocyanate from pesticide factory (Union Carbide).

Bio fuel – It is liquid or gaseous fuel made up of plant material.

Biodegradable – When substances get broken into simpler substances by decomposers.

Biodiversity – It is the variety of plants, animals and microorganisms on the Earth.

Biological oxygen demand – Dissolved oxygen needed to decompose organic materials by aerobic organisms in a given period of time and given volume of water at a specific temperature.

Biomass – Total dry matter of organisms.

Biome – Land ecosystems.

Biosphere – It is the environment which includes lithosphere, atmosphere, hydrosphere along with man, plants, animals and organisms.

Biotic – Living.

Carnivore – Animals that depend on other animals for their food.

Carrying capacity – Maximum population that an ecosystem can support over a given period of time.

CFC's – Chlorofluoro carbons used in refrigerators and air conditions as a coolant.

Chipko movement – Movement started in U.P. by Sunder Lal Bahuguna against the cutting of trees.

Climate – General pattern of atmospheric or seasonal variations over a long period.

Coal – Mixture of organic compounds, formed from the remains of plants when subjected at higher heat and pressure over millions of years.

Coastal zone – This zone is shallow part of the ocean, nutrient rich and warmer.

Commercial hunting – Hunting of wild animals for profit from sale.

Community – Population of different plants and animals living in an area at a particular time.

Competition – When two or more individuals of a single species or different species use the same resource in the same ecosystem.

Composting – Decomposition of organic wastes by the action of microorganisms.

Concentration – Chemicals in a specific volume or weight of water, soil and air.

Confined aquifer – It is the ground water between two layer of rocks having low permeability.

Coniferous trees – These are cone bearing trees.

Conservation – Protection, use and management of resource so that they are not depleted.

Consumer – Organisms that depend on other animals or plants for their food. They get their food by decomposition or by eating other organisms.

Consumption overpopulation – When people utilise resources at a higher rate and give rise to pollution, depletion and environmental degradation.

Contaminant – It is the material that is introduced by man activities into the environment.

Contraceptive – Method to prevent pregnancy.

Core – Inner zone of the solid earth.

Crop rotation – Plantation in same areas with different crops from year to year.

Crust – Outer zone of the solid Earth.

Deciduous trees – Trees that lose their leaves during some part of the year.

Decomposer – Organisms that breakdown organic matter into simpler forms.

Deforestation – Cutting of trees.

Desalination – Removal of dissolved salts from salt water.

Desert – Terrestrial ecosystem where evaporation exceeds precipitation.

Desertification – Process of converting cropland, rangeland into desert-like land.

Detritivores – These are consumer organisms that feed on organic matters.

Dissolved oxygen – It is the amount of oxygen dissolved in a certain amount of water at a specific temperature and pressure.

Diversity – Variety.

Drip irrigation – Method of irrigation by means of small tubes or pipes.

Drought – Area that lacks water.

Ecological diversity – Variety of ecosystems.

Ecological succession – It is the process when communities are replaced by series of different communities in a specified area.

Ecology – Branch of science that deals with the study of interactions of living organisms and the environment.

Effluent – It is a liquid coming from a factory or an industrial area.

Endangered species – Species that become extinct.

Energy – Ability to do work.

Environment – External conditions that causes affect on the organisms during its life.

Environmental degradation – Due to man-made activities the quality of environment gets depleted.

Environmental studies – Branch of science that deals with all aspects of human environment.

Estuarine zone – It is zone near coastline.

Estuary – It is the zone where river water mixes with sea water near coastline.

Ethics – What we believe to be wrong or right behaviour.

Euphotic zone – It is a zone of ocean where maximum penetration of sunlight takes place.

Eutrophication – Nutrient rich lake with low dissolved oxygen and high amount of algal blooms.

Extinction – Disappearance of species.

Family planning – It is a method of planning or spacing of children by using contraceptives.

Fertilizer – Substances that increase fertility of soil.

Fishing – Capturing of fish species.

Food chain – Series of organisms, each eating the preceeding one.

Food web – Complex network of food chains.

Forest – Terrestrial ecosystem.

Fungicides – Chemicals used to kill fungi.

Fungus – A eukaryotic, multicelled organism with chitinous cell wall.

Gamma rays – These are ionising, high energy radiations and penetrate into body.

Ganga action plan – Programme launched by government for cleaning up Ganga.

Genetic diversity – Variability in the genes within a species.

Geothermal energy – Production of electricity from Earth's core.

Global warming – Increase in the temperature of Earth due to green house effect.

Grassland ecosystem – Terrestrial ecosystem with the dominence of grass.

Ground water – Water resource located below the Earth surface.

Habitat – Area where an organisms live.

Hazard – Things that can deplete the environment and cause injury or disease.

Herbicides – Chemicals used to kill smalle plants or inhibit growth.

Herbivore – Organisms that feed on plants.

Humus – Partially decomposed complex mixture of organic and inorganic compounds.

Hydrological cycle – Natural cycle of water that distributes the water from Earth to the environment.

Hydrosphere – All types of water resources.

Hygiene – It is the branch of science related to health.

Hypolimnion – Bottom zone of water in stratification of lake.

Incineration – When solid wastes burns into ashes.

Infant mortality – Number of infant deaths per 1000.

Infra red radiations – Long wavelength radiations.

Insecticides – Chemicals used to kill insects.

Ionising radiations – High energy radiations that can damage living tissues.

Kelvin – Unit of temperature.

Kilo calorie – Unit of energy.

Kilo watt – Unit of electrical power.

Lake – Standing fresh water body.

Leaching – Different chemicals in upper layes of soil are dissolved to lower layer.

Less developed country – Country that have low industrialization and GNP per person.

Limnetic zone – Upper layer of lake, where maximum penetration of sunlight takes place.

Lithosphere – It is the Earth's crust consisting of rocks and soil.

Littoral zone – It is the shallow water, where penetration of sunlight takes place.

Malnutrition – Diet with faulty nutrition.

Mantle – It is the zone between core and crust of earth.

Mass transit – Transportation that carry large numbers of people.

Mature community – It is the advanced stage of ecological succession.

Mineral – Naturally occurring inorganic solid with specific chemicals.

Mono culture – Single crop cultivation.

More developed countries – Countries, that have high GNP per person and highly industrialised.

Natural gas – Underground deposits of gases.

Natural resources – Resources produced by Earth's natural processes.

Noise pollution – Noise above tolerable level.

Non biodegradable – Substances that cannot be breakdown into simpler substances by biological agents.

Non renewable resources – Resources available in fixed amount in Earth crust.

Nuclear fission – Splitting of uranium or plutonium into smaller fragments and release of large amount of energy.

Nuclear power – Generation of electricity from nuclear fission.

Ocean energy – Generation of electrical energy from oceanic waves.

Oligotrophic lake – Nutrient deficient lake.

Omnivore – Animals that feeds upon plants and animals as food.

Open sea – Area of ocean that is beyond the continental shelf.

Ore – Deposits having higher concentration of minerals.

Organic fertiliser – Organic materials like green manure, animal manure.

Organic pollutants – Pollutants that are organic in nature.

Over fishing – Harvesting of large number of fishes.

Over nutrition – Food with high calories.

Over population – Condition where life support systems are impaired.

Overgrazing – Consumption of grasses by animals to the point that its renewal is very difficult.

Ozone hole – Thinning of ozone layer by the interactions of chlorofluorocarbons, nitrogen oxides with ozone.

Ozone layer – Layer of atmosphere that filters the ultraviolet radiations from the sun.

Parasite – Organisms that feeds upon living organisms.

Permeability – Degree to which water can flow freely from rock pores that are interconnected.

Pesticide – Chemical used to kill the organism.

Phyto plankton – Small plants found in aquatic ecosystem.

Pioneer community – Initial community found in primary ecological succession area.

Point source – This is an identified source from where pollutants are discharged.

Pollution – Any undesirable change that affects health.

Population explosion – Population grows at a fast rate so that its size exceeds the carrying capacity.

Porosity – Pores in rocks or soil.

Poverty – Unability to meet the basic needs.

Predator – An organism that feeds and captures other organisms.

Prey – Organisms that serve as a source of food for an organism.

Primary pollutant – Pollutants which are primary in nature such as carbon monoxide, hydrocarbons etc.

Producer – Organisms that manufacture their own food by the use of solar energy.

Profundal zone – This zone is deep and present in a lake. Here, sunlight is not penetrated.

Radioactive pollution – Pollution that occurs due to radioactive materials.

Radioactive wastes – Wastes generated from nuclear power plants, weapons, medicines, and research involving nuclear reactions.

Rangeland – Land that provides food in the form plants for grazing animals.

Recharge area – Zone in which an aquifer is replenished with water by the percolation of precipitation through rock and soil.

Recycling – Reprocessing of resource.

Renewable energy resource – Energy which can be renewed such as tidal energy and solar energy etc.

Renewable resource – Resource which can be renewed and used again such as grasses.

Reservoir – It is built behind a dam and human created standing fresh water body.

Resource – Anything obtained from the environment to fulfill the human needs.

Reuse – Use of a product again and again in the same form.

Rock – Aggregation of minerals.

Runoff – Fresh water from melting ice or precipitation flows on the Earth's surface into lakes, streams etc.

Salinity – Amount of dissolved salt in a given volume of water.

Salinization – Accumulation of salts in soil that prevent growth of plants.

Sancturies – Area reserved for wild animals.

Sanitary land fill – Solid wastes are disposed off in a pit covered with soil layers.

Secondary air pollutant – Chemicals formed after the reaction with normal air pollutants.

Sediment insoluble – Solids settled in water body at the bottom.

Shifting cultivation – Cutting down trees in a forest for planting crops for a few years.

Silent valley movement – Environmental movement organised by NGO Kerala.

Sink – Medium that retains a pollutant.

Soil – Mixture of inorganic minerals, organic matter, air, living organisms and water.

Solid waste – Unwanted solid material.

Species – Organisms of the same kind.

Species diversity – Variety of a number of species.

Sport hunting – Killing of animals for recreation.

Stocholm conference – First United Nations conferences in 1972 on human and environment held at Stocholm.

Stratosphere – Second zone of atmosphere present about 19 to 48 kilometer from Earth's surface.

Strip cropping – Planting close growing crops.

Strip mining – Mining is done with bulldozers to remove large chunks of Earth's surface in strips.

Subsurface mining – Extraction of mineral deposits from deep underground deposits.

Surface mining – Extraction of minerals deposits close to the Earth's surface.

Surface water – Precipitation that does not infilterate into the soil.

Terracing – Planting crops on terraces that retain the slope of water.

Tertiary consumers – Organisms that feed on animal eating animals.

Thermal pollution – Rise of water temperature resulting in harmful effects on an aquatic ecosystem.

Threatened species – Species that is abundant in nature but due to killing of species is likely to become endangered.

Tidal energy – Tidal waves are converted into electrical energy.

Toxic substances – Chemicals harmful to organisms.

Toxic waste – Waste that causes death to organisms.

Transmissible disease – Disease caused by living organism, that spreads from one person to another person.

Trophic level – Organisms that consume the same type of food in a food chain.

Unconfined aquifer – Collection of groundwater above a layer of rock through which water flows slowly.

Under nutrition – Taking in less food for a longer time, causing harmful effects.

Waste land – Uncultivated land.

Water consumption – Water withdrawn from groundwater and is not available for reuse in that area due to transpiration or evaporation.

Water logging – Saturation of soil with irrigation water and rise of water table to the surface.

Water pollution – Any undesirable change in ground water or surface water that causes harmful effects.

Water table – Layer of soil surface in which all the pores are filled with water.

Water withdrawl – Removal of water from ground water and it's transportation to for use.

Watershed – Land area that provides runoff water or sediments to the bodies of surface water.

Wetland – Land flooded with fresh or salt water throughout the year or partly.

Wildlife – Undomesticated species.

Wind energy – Strong winds that are converted into electrical energy.

Zone of saturation – Land area where all the pores of soil and rock are filled with water.

Zooplanktons – Small floating herbivores that feed on phytoplanktons.